Anemonefish (*Amphiprion* sp.)

Penguins on the coast near the Ross Sea, Antarctica

Weddell seal *(Leptonychotes weddellii)* in its breathing hole

Can man learn to live as part of the ocean world?

Sea elephants greet the dawn on San Miguel Island, California

Harbor at Wanchese, North Carolina, with excursion fishing
boats moored for the night

The dugout canoe and the net, two of the earliest inventions, survive today in many fishing cultures

THE DRAMA OF THE
OCEANS

Elisabeth Mann Borgese

THE DRAMA OF THE

OCEANS

Harry N. Abrams, Inc., Publishers, New York

Library of Congress Cataloging in Publication Data

Borgese, Elisabeth Mann.
    The drama of the oceans.

    Bibliography: p.
    1. Marine biology.   2. Ocean.   I. Title.
QH91.B66   551.4′6   74-16165
ISBN 0-8109-0337-7

Library of Congress Catalogue Card Number: 74-16165
Published by Harry N. Abrams, Incorporated, New York, 1975
Printed and bound in Japan

The lines quoted from Robert Fitzgerald's translation of Homer's *The Odyssey*
are reprinted by permission of Doubleday & Co., Inc.

CONTENTS

ACKNOWLEDGMENTS

I am indebted to all the companions in the struggle for Peace in the Oceans
(Pacem in Maribus), and to all members and associates of the International Ocean
Institute, Malta, who sent pictures and contributed ideas and inspiration.
In particular, my thanks go to Professor John Bardach, University of Hawaii;
Mr. Thomas Busha, Intergovernmental Maritime Consultative Organization
(IMCO); Professor Roger Charlier, Northeast Illinois University and University
of Bordeaux; Dr. Peter Dohrn, Aquarium, Naples; Mr. Clifton Fadiman,
Santa Barbara, California; Mr. Thomas Gaskell, British Petroleum, London;
Dr. Thor Heyerdahl, Liguria, Italy; Dr. Sidney Holt, Food and Agriculture
Organization (FAO); Professor William Murdoch, University of California;
Dr. Arvid Pardo, Malta; Dr. Jacques Piccard, Switzerland; and Professor Roger
Revelle, Harvard University, who read the manuscript, gave invaluable criticism,
made most constructive suggestions, and provided precious source material; to
Shirley Henderson, who helped with the research for chapters 1 through 4; Stephen
Sprinkel, who did much work on chapter 4, and David Krieger, for his contribu-
tions to chapters 8 and 10; to Sam Scranton and Cynthia Travis, who helped with
the pictures, and to Jean de Muller, who helped throughout. It has been a
pleasure to work with the editorial staff of Harry N. Abrams, Inc., and my
thanks go particularly to Mr. Harry Abrams personally and to Dr. Fritz H.
Landshoff for their encouragement, and to Mrs. Margaret L. Kaplan and
Mrs. Lisa Pontell for their patient, intelligent, and indefatigable collaboration.
Mr. Nai Y. Chang created the elegant design.

THE DRAMA OF THE
OCEANS

Free man, you always will cherish the sea!
The sea is your mirror, you study your soul
In the infinite roll of its billows,
And your spirit is a gulf no less bitter.

Baudelaire, *The Flowers of Evil*

PROLOGUE

WHY? WHY ARE THE OCEANS THE MIRROR OF OUR SOUL, more than mountains, more than deserts, more than forests or the firmaments set in their eternal rhythms?

The oceans contain them all. Mountains—those that the surfer experiences, more overwhelming than the skier's mountains; and those which science is but beginning to fathom: ridges and craggy peaks, rising from the seabed, that dwarf the Himalayas; abysses that could swallow Mount Everest and Mont Blanc piled one upon the other; volcanoes spewing their red-hot lava through the blue waters; deserts of red clay stretching farther than ten Saharas; beds of sand; and submarine meadows and forests with a wealth of species of flora and fauna, ageless and endless, that makes life on land appear as a fleeting episode.

At the beginning of time there was water, only water in the starless night of the lifeless interval between dissolution and creation. All the potentialities of subsequent evolution rested, dormant and undifferentiated, in the primeval sea.

The Egyptians called it Nun, the ocean or shapeless magma containing all the seeds of life.

The earth may have floated in the midst of it like the yolk of an egg, if the Chinese were right. Or a magnificent lotus may have grown on it, and a Divine Being may have risen from the depth through the stem of the lotus, as the Indians have it. If this version is correct, He ordered some animals to bring Him mud from the bottom of the sea, and with this He fashioned the earth.

Or perhaps Izanagi and Izanami came forward together over the floating bridge of heaven—just over Japan—and plunged a jewel-bedecked celestial spear into the ocean of chaos that stretched beneath them. They stirred it until the liquid coagulated and thickened. When they withdrew the spear, drops of brine fell from it and formed the beginning of the world, an island "that coagulated of its own accord."

The primeval waters must have been dismal, and some of their dismal features persisted in the oceans after the world was created. Thus the Arctic people maintain that the sea is a vast river. To set sail on the river leads to the village of the dead. And the death of the ocean will overtake us again at the end of time, if there is an end, for the cycle may repeat itself endlessly, destruction alternating with creation.

That the failure of man and divine wrath brought on torrential rains and tidal floods that engulfed the earth was known to the people of Israel as of Peru, of Babylon as of Japan and India.

"Oh King of Gods," Brahma said to Vishnu, "I have seen all perish, again and again, at the end of every cycle. At that time, every single atom dissolves into the primeval water of eternity, whence originally all life arose. Everything then goes back to the fathomless, wild infinity of the ocean which is covered with utter darkness and is empty of every sign of animate being."

In the West we have inherited Israel's version of the Creation, God dividing light from darkness and earth from the waters. But lately we have done some deeper probing into the question of where the water came from in the first place.

There are numerous versions.

The earth may have been enveloped in heavy layers of clouds, so thick that no ray of sunlight could penetrate. In the darkness of this cloud cover, the earth's surface cooled, and when it was cool enough to receive moisture without sending it steaming back into the cloud cover, then the rain began to fall, and it rained without surcease for thousands, perhaps millions of years. This is supposed to have happened about two and a half billion years ago.

From the moment the rain began to fall, the land began to erode. Rocks dissolved, and their minerals were carried into the expanding sea, increasing its salinity over the eons.

A later version suggests that the water was distilled from the earth's interior by volcanoes. As the nascent planet whirled through space, its heavy metals were attracted to the center of the molten core, and the lighter components, including water, were centrifuged to the outside through volcanic valves. The creation of basins, in which the planet's waters drained, is an evolutionary, continuous process, with molten earth crust extruding from mid-ocean ridges and forcing the continents apart or into mountain-raising collisions.

All versions agree that every drop of water that existed on the earth or around it billions of years ago is still there, whether in solid form or liquid or gaseous, on the surface or deep underground, free or encapsuled in rock or living creature. Every drop is still there.

These accounts are science, of course, rather than myth, but they are no less beautiful for that.

Every woman's womb is a micro-ocean, the salinity of its fluid resembling that of primeval waters; and every micro-ocean restages the drama of the origin of life in the gestation of every embryo, from one-cell protozoan through all the phases of gill-breathing and amphibian, to mammalian evolution.

And every human, in turn, is a planet ocean, for 71 percent of his substance consists of salty water, just as 71 percent of the earth is covered by the oceans.

Now imagine that every day in the gestation of a human grew into an eternity, and that every embryonic moment gave rise to an evolution of its own, and that all the species thus evolved were still with us.

The world's oceans contain them almost all, in unbroken continuity. And just as the past lives in the oceans' timelessness, so does the future, and marine life embraces the return of the avian as of the mammal to the world womb; and wings, legs, and arms become flippers again, whether of cartilage and skin or of rubber and plastic.

The eternity of the oceans is whiteness which contains all the moments of the rainbow of time. And the noise of the sea, that eternal roar which mutes and deadens the sense of time, is white noise, containing the rhythms of the planetary system and the drumming of catfish, the whistle of dolphins, the keening of whales.

Beneath the great shroud of the sea, all seems to be silence as well as darkness. But though the darkness is pierced here and there by bioluminescence, "sight" in the ocean depth is chiefly by sound and smell.

The wide-open spaces of the sea are a dense network of communications on which the survival of all life depends. Simple marine bacteria, those tiny organisms which are responsible for the recycling of organic matter in the oceans, communicate with each other by chemical signals. Octopuses speak to one another by changing the color of their skin. Salmon and herring seem to be guided on their long journeys by distinct smells. Dolphins have an echo-sounding capacity that enables them to identify the distance, shape, substance, and texture of things in the sea. Aggressiveness and cooperation, lust and acquiescence, in lobster and catfish, in goby and crab, are responses to chemical signals that blend into "white smell"—the briny tang that we call sea air.

But this ocean system is as delicate as it is vast, and as much as we may think we know about it, the gulf of our ignorance is deeper than the deep ocean trenches. Man's foolishness and greed are seen by science, as by myth, as eventually bringing back the lifelessness of the primeval waters.

Reckless tampering with the macrotechnologies of the twentieth and twenty-first centuries may melt the polar ice cap, raise the level of the oceans enough to engulf coastal cities, or perhaps bring on a new ice age. Marine exploration and exploitation may disrupt delicate systems of communication and the cycles of reproduction based on them, and many species may die out. The industrial effluents and domestic sewage of growing populations, and the radioactive wastes generated by succeeding phases of the arms race, may eventually kill the phytoplankton in the upper layers of the sea, which produces more than half of the world's oxygen. Should this phytoplankton disappear, all life in the oceans would cease. Oxygen shortage and the stench of putrefaction would send coastal populations scrambling toward interior, mountainous areas. Many would perish in the struggle for the decreasing supply of breathable air. The rest would die asphyxiated when the oxygen supply was exhausted over the wild infinity of the ocean, empty of every sign of animate being.

This would be one ending of the drama of the oceans. But if man is indeed free he need not accept such a fate.

Hermann Muller, Nobel-laureate biologist, wrote: "Through billions of years of blind mutation, microbes finally emerged as man. We are no longer blind; at last we are beginning to be conscious of what has happened and of what may happen. From now on, evolution is what we make it."

If we can remake ourselves in our environment—both social and natural, the one being part of the other—the oceans will live, the oceans will be bountiful. And they will mirror the souls of a free mankind.

PART ONE

THE SCENE

... there is some encouragement in the reflection that Oceanography has usually only ruined the reputations of people who dared to speculate too little and thought on too small a scale. She has smiled most benignly on those who backed the most daring and outrageous possibility.

Brenda Horsfield and Peter Bennet Stone,
The Great Ocean Business

CHAPTER ONE
THE GEOPHYSICAL SETTING

IMAGINE THAT WE ARE FLYING OVER THE ALPS or the Rockies or the Himalayas, and that the deep-blue sky is carpeted with clouds, gray and blackish, gently billowing and rippling—an ocean of clouds with a few mountain peaks, snow-covered, jutting up like islands. Somehow it feels as though we are sailing on these stormy clouds in a ship. Then the wind dissipates the clouds. As they drift away, an awesome panorama expands under our wings—ridges, valleys, gorges, peaks and crests, plateaus, cliffs, rocks, greenery, and snow—all losing themselves in the blue distance.

The vision conveys a sense of motion. While rationally we think of "floor," of "rest," of "permanence," intuitively we think of waves, of billows—whether foam-crested or snow-capped—rising high toward the sky, to break, to come crashing down.

The only difference between our rational and our intuitive thought is made by time. In human time these frozen waves are eternal. In geological time, in which a year is no more than the duration of a heartbeat, we would see mountain ranges being thrown up by the clash of continental masses, the Himalayas piling up when Hindustan and Asia collided, the Alps amassing when the Italian peninsula ran into Europe.

Suppose we had filmed it and could rerun a few hundred million years in the moment it takes a child to squeeze two handfuls of clay together—clay with a crest thrown up at the line of greatest pressure, with billions of microorganisms churned around and displaced. The picture of the two events would be much the same. But whose time is real time—the microorganism's, man's, or the planet's?

Looking at the ocean, the opaqueness of the rippling surface seems like the cloud cover over which we flew. Islands rise above, like peaks and high plateaus.

Many people by now have dived into the sea to explore the hidden landscape. "It's like flying a helicopter in the Rocky Mountains on a dark night," as one diver put it, and the night is not only dark, it is foggy as well—a pea-soup kind of fog almost impenetrable to light. If a light is flashed ahead of a submarine, the effect is like that of a car's headlights in fog: so much light is scattered back that, except at very close range, the situation is worse than before.

Yet the fog can be dispelled. We can see, if not through the porthole, then on the submarine's television screen. We have the technology to see beyond the fog, beyond the darkness: sonar (echo sounding); laser technology (sounding with light beams so concentrated that they can cut through not only water but even diamonds); computer science and holography (the three-dimensional reconstruction of three-dimensional space, whether drawn by ultrasound or by light). All these let us view new horizons, a world far larger than the one we know, more challenging than the world of outer space. The awareness of this world is beginning to revolutionize our concept of the nature and history of the planet earth, the origin and end of all things, man's common heritage in the most literal sense.

Let us project our technological capabilities a little beyond what they are now. Let us imagine a television screen wide enough to give us the kind of view we saw when flying over the Alps, a screen showing a three-dimensional picture and a color scale based on the density of the materials scanned. Computers can put together such images. Now imagine that we are traveling in a comfortable submersible, of the type of the *Aluminaut* or the *Ben Franklin,* down the Atlantic and up the Pacific, looking at our television screen.

We start from Iceland, for Iceland is one of the formations of the underwater world that rise up through the water cover.

Iceland is made of different material than that of continents. It is made of basalt that poured out of volcanoes and formed sheets that average thirty feet in thickness. Granite, the bedrock of continents, is totally absent.

The volcanoes are still there and are still pouring, although the area of their activity is shrinking and the rhythm of their outbreaks is slowing. In the postglacial period, about 200 volcanoes were active. Since the first settlers arrived in Iceland in the ninth century A.D., about thirty volcanoes have been known to be active, and during the past few centuries there has been an eruption once every five or six years. In 1963 a violent one ten miles southwest of Vestmannaeyjar (the Westman Islands) sent red spurts of fire through the icy blue, giving birth to a new island, Surtsey.

More lava has welled up from tiny Iceland's volcanoes during recorded history than all the lava of all the continents' volcanoes put together. Iceland has all types of volcanoes. There are crater rows, which resulted from single eruptions from linear vents or fissures forming chains of craters. There are also stratovolcanoes of the Mount Fuji type, some of them ice-capped. Their eruptions are accompanied by tremendous floods, or glacier bursts, whose discharge may exceed that of the Amazon River. The most famous volcano, Hekla, was created by numerous fissure eruptions that formed a vaulted ridge reaching a maximum height of 4,747 feet. During eruptions the ridge opens lengthwise, forming a 3.5-mile fissure.

Here, on dry land, we can see what the mid-ocean ridge looks like. Sheer basalt rock. A deep rift that keeps expanding, breaking up the floor and tilting it into long, narrow, upturned blocks. A chain of volcanoes in the background.

From Iceland we follow the submersed part of the mountain range. The awesome spectacle of the Alps is puny by comparison. The Alps measure 680 miles in length and 80 to 140 miles in width. The highest peak, Mont Blanc, rises 15,781 feet from a ground elevation of 3,400 feet. But there is almost 40,000 miles of mid-ocean ridge system, and it averages 1,250 miles in width. The peaks rise 15,000 feet from the ocean floor.

We follow the Mid-Atlantic Ridge the whole length of the Atlantic. It runs north-south right through the middle, imitating the curves of the coastlines from the Central American concavity, into which the African convexity between Mauritania and Liberia would fit so neatly if only we could push them together, to the African concavity of Nigeria and the Cameroon coast, which the tip of western Brazil would fill. Mt. Jupiter. Mt. Venus. Mt. Pluto.

At the southern end the ridge system bends around the Cape of Good Hope into the Indian Ocean. There is a great deal of movement in the ridge line in this region, just as there is in the adjacent coastlines of southwest Africa on the one hand and the eastern coasts and islands of the Indian Ocean on the other.

The Southwest Indian Ridge divides the Madagascar Basin from the Crozet Basin, runs into the Carlsberg Ridge, pushes into the Gulf of Aden, and from there moves at a right angle into the Red Sea.

Toward the south, the Mid-Indian Ridge extends into the Southeast Indian Rise, which runs into the Pacific-Antarctic Ridge south of Australia, then moves parallel to the western coasts of Chile and Peru as the Albatross Cordillera (East Pacific Rise), and into the North American continent just below Baja California.

We seem to have arrived on another planet—a planet twice as large as earth, judging by the dimensions of this gigantic panorama. Its texture and shape are different—it is made of sheer volcanic basalt, and there are no such mountains on earth.

The ridge system's rugged crest is about 125 miles wide and lies at some points only about 10,000 feet below sea level.

Right through the middle of the entire ridge system runs a valley, eight miles wide at its narrowest, 130 miles at its widest. The Grand Canyon, one mile deep, four to eighteen miles wide, and 280 miles long, is modest in comparison.

Every forty miles or so the ridge is cut almost at right angles by a fracture line, or canyon, extending for thousands of miles on both sides of the ridge. With astounding regularity, each block of the ridge is displaced slightly to one side.

The crest is covered by strange glass-encrusted pillows that are formed as the hot lava, welling up into the floor of the rift valley, is suddenly quenched by the chilling waters. Let us listen to two scientists, Bruce Heezen and Charles Hollister, of the Woods Hole Oceanographic Institution, as we gaze on the unearthly surface of this panorama: "These wrinkled, fractured flows, only lightly dusted by ooze, were formed at a recent yesterday and are still measurably warm." The crests are adorned by sea pens, fans, sponges, and gorgonians. The steep slopes, like towering walls rising at angles of 60 or 70 degrees—toothpick formations jutting from the seabed—are often covered with shattered fragments of pillow lava. These, our guides say, "are characteristic of the walls of the Rift Valley." We can even detect a certain number of wedge-shaped fragments that "reveal the characteristic radial jointing of the pillows."

In the North Atlantic we turn off at a right angle, following the Atlantic Fracture Zone, and head west. Around us lies a landscape of seamounts, submarine mountains rising three miles above the abyssal plain. The plain is smooth and even, covered with fine sediments such as red clay or with an accumulation of minute animal skeletons, the radiolarians, which represent a record of life on this planet some 70 million years ago.

The seamounts jut up incredibly steep and ragged, like the worn teeth of a gigantic carnivore. They reach to about 120 feet below sea level and are capped by coral reef and sands. Sometimes they break the surface of the water and appear as islands, as, for instance, in Bermuda. The Bromley Plateau (Rio Grande Rise) off the Argentine coast offers similar phenomena.

There is more abyssal plain after we cross the seamounts, smooth and even, it covers half the planet's surface and is its flattest plain. Then the territory begins to rise. Gently at first, then more steeply. We have arrived at the continental rise. This softly rising floor, 145 miles wide on the average, links the steep cliffs of the continental slope with the deep ocean basin. Sediments and oozes increase as we move

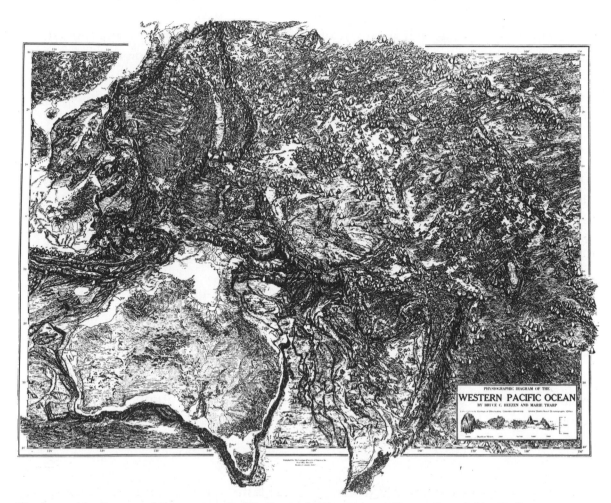

Physiographic diagram of the western Pacific Ocean by Heezen and Tharp. Copyright 1971 by Bruce C. Heezen. Published by the Geological Society of America, Inc., Boulder, Colorado. Reproduced by permission.

from the mid-ocean ridge toward the continent. There is red clay of a type unknown on terra firma. It is, in fact, partly of unearthly origin, containing meteoric iron, nickel, and silicon, of which eight to ten tons arrive from outer space every year. The oozes, however, are organic, made up of "snowfalls" of tiny skeletons and shells: the foraminifera, the diatoms, the radiolarians.

Light increases as we move shoreward. So does life. According to some estimates, the mid-ocean abyssal plain is inhabited by barely a milligram of living matter per square mile. When we reach the continental shelf, on an equivalent bottom area we find as much as 200 grams, while the inshore sea floor teems with living organisms to a density of 5,000 grams per square mile.

Delicate sponges as well as fans, feathers, anemones, coelenterates, and stalked sea pens grow in the abysses. Gorgonians display a variety of forms, some of them brightly luminescent. Heezen and Hollister describe one as shrublike, glowing with a soft pale lilac light in the abyssal night. Sea lilies, stars, urchins, and cucumbers—the echinoderms—are a dominant part of the seascape at all latitudes and depths. Feather gardens, mosses, and meadows become denser as we move landward.

We have arrived at the foot of the continental slope. The landscape is varied. Sometimes the ascent from the ocean depths is by steps, with marginal plateaus breaking the steepness of the slopes. At other places the gradient is quite steep, as off the west coast of Florida, where it looks like the scarp on the east side of the Sierra Nevada.

But we shall see yet more dramatic slopes. On the other side, on the Pacific coast of Chile, we have an almost vertical cliff slope nearly twice as high as the southern wall of the Himalayas, rising from the depths of the Peru-Chile Trench.

The deep trenches—most of which are in the Pacific—are the deepest depressions on the planet. The longest is the 3,660-mile Peru-Chile Trench. The Marianas Trench, curving for 1,250 miles along the Marianas Islands toward the southern end of the trench system whose northern end is formed by the Kuril-Kamchatka Trench, reaches a depth of 36,198 feet. This depth is the greatest ever attained by human beings—Jacques Piccard and Don Walsh in the bathyscaphe *Trieste* in 1960. Here is how Piccard described it to me:

I saw a wide circle brightly lit just below us. This was our projector light, which preceded us. And the *Trieste* continued to float, descending gently at a velocity of not more than a few centimeters per second. The light circle shrank slowly as we approached the bottom.

The bottom was covered by a sediment, light in substance and color—a vast desert, ivory-colored. Obviously we had landed on a mass of diatoms, sufficiently resistant for the guide rope to rest on. For a moment, well poised on this cable which weighs a mere 25 kilograms, the heavy mass of the *Trieste* was floating between two layers of water; then, slowly, as its fuel cooled down, the 150 tons we displaced and which now weighed only a few kilograms completed the conquest of these profound depths. At 13:06 hours exactly the *Trieste* came to rest on the bottom of the Marianas Trench, at a depth of almost 36,200 feet. At the moment of touchdown, a light cloud rose around us, enabling us to appraise even better the delicacy of this extraordinary sediment. None of the little hills so typical of smaller and medium depth were visible here, even though there was here and there some small irregularity in the soil. But above all, at the moment of our landing we had the immense good luck to see, just at the middle of the light circle thrown by our projector, a fish. Thus, in one second—even though after years of preparations—we could answer the question which thousands of oceanographers had been asking for decades. Higher forms of life *do* occur anywhere in the ocean, no matter what the depth. What we saw was evidently a true bony fish, very much like a sole, about thirty centimeters long and fifteen centimeters wide.

But let us return to the slope. Often it is carved by canyons running like river valleys for lengths ranging from a few miles to several hundred. Some are a few hundred yards wide, others several miles. They may be just a few hundred yards deep, but some reach a depth of more than a mile. The walls are steep, often nearly vertical—greenish brownish, blackish. The Congo Canyon, cutting through the West African continental slope, has 6,000-foot-high sheer cliff walls on both sides.

Once we are up the slope, we have passed the border between darkness and light. There is a sharp break in gradient, and we find ourselves on the continental shelf.

The continental shelf is the submerged extension of the continent. It descends slowly, over a width ranging from a few miles to 750 miles, from the low-water line to a depth of 1,200 to 1,800 feet. The sum of all continental shelf areas in the world measures only about one-tenth of the surface of the oceans, but this is where most of the activity is, where the land and sea interact.

We pass over sandy shelves, submerged rippling beaches. We pass over rocky shelves and caves of limestone or granite, the rock of which continents are made. And we pass over lush vegetation—fields of algae, red, green, and brown, from tiny plankton to tree size; rockweeds and kelp, sea pastures of Zostera, turtle and eel grasses; coral gardens; sea cucumbers and anemones. All sorts of creatures, including man, weave in and out.

We have cheated time. At a glance we have seen the results of thousands upon thousands of discoveries, measurements, projections, scientific processes of great variety and ingenuity developed over the last decades—decades of discovery that have been compared to the era of Columbus, Vespucci, and Magellan.

Now suppose that in viewing our screen we decide to make our computer reconstruct the sequences—running over hundreds of millions of years—that the ocean depths have recorded for us. What a film that would be!

What began as the stage setting for our drama of the ocean turns out to be a drama in its own right. For there is a drama within a drama within a drama: man's multiple activities and interactions with the world ocean, set within the evolution of life from the ocean and returning to the ocean, set within the eternal transformation of the planet itself.

The first insight into the larger drama of the ocean was religious. The Book of Genesis relates that it was Noah's flood that divided the continents, which until then had been united. The first to expand on this revelation was a French theologian, François Placet. In his book *La Corruption du grand et petit monde,* published in 1666, he holds that "before the Deluge America was not separated from the other parts of the Earth, and there were no islands."

Ninety years later another theologian, Theodor Christoph Lilienthal of the University of Königsberg, confirmed that it was indeed Noah's flood that had divided the continents; prior to that event, the continents were united, as manifested by the fact that "the facing coasts of many countries, though separated by the sea, have a congruent shape, so that they would almost fill one another if they stood side by side: for example, the southern parts of America and Africa."

With Lilienthal we have already passed from religious intuition to scientific intuition—scientific insight that could not yet be corroborated by the tools and methodologies at hand. The most daring of these intuitions came from Alfred Wegener, a German meteorologist and explorer who first proposed his theory in 1912. Like his predecessors, Wegener was puzzled by the fit of continental contours. His curiosity was sharpened by fossil evidence of inexplicable migrations of species between continents separated by wide oceans. Basing his arguments on previous findings by imaginative Germans such as Baron Carl Löffelholz von Colberg and H. Wettstein, he adduced geophysical, geological, paleontological, and biological as well as geodetic evidence in favor of his theory, which certainly did not lack grandiosity. "One thing may be accepted as certain," Wegener wrote. "Continental drift, folding and rifting, vulcanicity, the alternation of transgressions, and the wanderings of the poles, stand in one great causal connection with one another" (*The Origins of Continents and Oceans*).

Unlike his predecessors, who postulated one great catastrophe—Noah's flood or some similar event—as the cause of the separation of continents, Wegener proposed a drift that started somehow in Jurassic time and is still going on. He assumed that originally the South American continental plateau directly adjoined the African plateau, forming one large connected mass. In Cretaceous time this mass split in two and the two parts, like floating icebergs, drifted farther and farther apart. Similarly, North America was close to Europe and formed with Greenland (at least from Newfoundland and Ireland northward) one connected block, which broke up with a forked rift near Greenland at the end of Tertiary time and farther north in the Quaternary era. Thereupon the constituent blocks moved apart from one another.

Antarctica, Australia, and India adjoined South Africa. Until the beginning of the Jurassic period, they formed a single large continental area, which also included South America—though parts of it were at times submerged by shallow water. In the course of Jurassic, Cretaceous, and Tertiary time this continental mass split and crumbled into smaller blocks, which drifted away from each other in all directions.

Wegener's unusual theories met with little success when he first proposed them. It took nearly half a century for him to be vindicated.

New techniques—drilling and coring, echo sounding, radioactive dating, rock-magnetic measurements—produced theories of a magnitude even more grandiose than Wegener's about the genesis of oceans and continents.

The architects of these theories came mostly from the United States, Canada, and Great Britain, and did their main work in the 1950s and 1960s. The most important of them were, perhaps, Harry Hess of Princeton, who published his theories in an essay, "History of Ocean Basins," in 1962, his collaborator, Robert S. Dietz, and John Tuzo Wilson of the University of Toronto, whose path-breaking paper on "transform faults" was published in 1965.

According to this new branch of science, oceans and continents are created and re-created in a continuous process. "The whole ocean is virtually swept clean every 300 to 400 million years," Hess wrote. The drama begins in the middle of the ocean, in the deep rift that halves the mid-ocean ridges and from which molten basalt pours forth, enlarging the earth's crust on a spreading ocean floor. During a human lifetime the sea floor moves, roughly speaking, the length of a human body. Thus, in a way, the oceans grow as continents are forced apart. The continents, whose granitic rock is lighter than that of the basaltic ocean floor, are forced apart, sliding on vast tectonic plates on the basalt ground. At present there are seven major plates, some without continents on them. They separate, they clash. When continents clash, mountains rise, pushing submarine surfaces high up into the sky. Hence, we find sea-shell fossils in the rocks of the Himalayas and reliefs of fish impressed into the walls of the Matterhorn. Ammonites, whose shells abundantly adorn the walls of the high north ranges of the Himalayas, were still living in the seas at the time the Himalayas arose.

We can trace this odyssey of continents for about 180 million years. Computer models, based on data furnished by deep-sea drilling, satellite photography, and other methods and technologies, have confirmed Alfred Wegener's vision of Pangaea, the supercontinent, in which 180 million years ago all our present continents were lumped together.

Pangaea must have biforked, like a pair of scissors, opening toward the east with the two blades converging on a juncture between what is now southern Europe and southwest Africa. The southern blade of Pangaea was Gondwanaland, consisting of what is now Antarctica, Australia, and the Indian subcontinent; the northern blade was Laurasia, consisting of Asia without the Indian subcontinent. The scissors' handle was formed by the Americas, fitting the contours of western Africa and Europe. Between the eastward-opening scissors was an ocean, the Tethys Sea.

About 190 million years ago, the Atlantic began to open in the Gulf of Mexico, and what is contemporary North America and Africa began to part. As the Atlantic rift expanded, South America, too, moved away from Africa, about 135 million years ago, and Antarctica, still joined to Australia, detached itself from the south. The separation of North America from Europe came later, followed yet more recently by the splitting of Antarctica from Australia. The Indian subcontinent traversed—and destroyed—the Tethys Sea, opening the Indian Ocean in its wake, about 90 million years ago and ran into Asia about 50 million years ago. So violent was the collision that it caused the Himalayas to splash high heavenward. And Hindustan, it seems, is still on the move.

North America is still moving away from Europe, and South America from western Africa. California is apparently approaching Alaska, and France and Italy will merge with North Africa. As North America slides away from Europe, it moves toward Asia. In this shifting of land masses some oceans, such as the Atlantic, are expanding; others, such as the Pacific and the Mediterranean Sea, are slowly disappearing, like the Tethys Sea. The expanding oceans do so from the center, their floors spreading from the magma-spewing mid-ocean rift; the shrinking oceans shrink from the margins, where a system of deep trenches devours ocean floor faster than it is produced at the center. It sinks into these maws; the earth quakes, and its molten core slaps over the rims of volcanoes all along island arches that accompany the deep trenches.

Every 300 to 400 million years the earth is made over, inside out and outside in, like a kneaded yeast dough. Only the continents, which are too light to be sucked into the trenches, keep floating outside on their tectonic plates, like bits of foam, merging, separating, transforming, turning around, changing climes, flora, and fauna.

It is always surprising to what extent scientific findings and theories confirm mythological traditions. We know now that the Great Flood took place many times, and that dragons very much like those depicted in fairy tales really lived—and in fact some, such as the giant lizard in Indonesia, still do live. We also know that Atlantis did exist, a Mediterranean island off the southern coast of Greece, ruled by a city with 10,000 chariots and 1,200 ships. Thera was its ancient name; its remains are known today as Thera or Santorini. As reconstructed by archaeologists and marine geologists, Thera's partial extinction came about through a series of disasters. Volcanic eruptions buried the city in lava and ashes, just as Herculaneum and Pompeii were later buried by Vesuvius. Then, between 1500 and 1400 B.C., the island was blown up by another volcanic explosion and much of it sank into the sea. Even in our own time such disasters have taken place. The island of Krakatoa in Indonesia exploded in 1883 and was engulfed by the sea. But measurements suggest that the explosion of Thera was about five times more violent than Krakatoa's. Ancient nature matched the destructiveness of our own modern atomic technology. A huge mushroom cloud rose from Thera, carrying poisonous vapors and dust to nearby Crete, whose Bronze Age civilization it destroyed, and as far as Egypt, which it covered with thick darkness for three days. Huge waves—tsunamis—were unleashed by the seaquake that accompanied the explosion, drowning coastal settlements far and wide.

As we now know, it was not the hubris of the trading city or the wrath of the gods that wrought this destruction. The catastrophe of Thera was only one episode in the slow but sure closing of the Mediterranean Sea. The Aegean plate, carrying southern Greece, Crete, and the Aegean Isthmus, is moving southeastward toward Africa. There is an ocean-floor-devouring trench off the coast of Crete, where the African plate is consumed; the Balkans and Asia Minor and the Riviera and North Africa will all meet, and a mountain chain more majestic than the Alps will replace the Mediterranean Sea. "Waves" of frozen motion, masses clashing, billows— whether foam-crested or snow-capped—rising high toward the sky, to break, to come crashing down.

Science or myth? To visitors from another planet—or even another culture—the line of demarcation between one and the other might be far less sharp than it seems to us. What appears certain, however, is that we have penetrated deeper into ocean space, physically and intellectually, than any generation before us.

There is another remarkable feature of this conquest.

Ocean research, as we have seen, draws on a great many disciplines: geology, geography, paleontology, physics, chemistry, acoustics, laser technology, holography, magnetology, ecology, and all branches of biology, as well as mythology and cosmology.

A lengthening series of the major problems of the contemporary world are of just such an interdisciplinary nature: from the environment to development, from energy to demography. But no other area has manifested its interdisciplinary nature so clearly as has that of marine studies. No other field of study has so inevitably forced the concept of interdisciplinary cooperation on our minds.

And just as it is interdisciplinary, so is it, perforce, international. Ocean science is too big to be handled by any one nation. The world ocean does not belong to any nation, and if it is to yield its secrets it must be studied in its entirety. Ocean research, furthermore, is too expensive to permit excessive waste and duplication of effort. Small nations cannot afford to undertake it single-handed. During its great revolution over the past twenty years, ocean research has thus become an increasingly cooperative effort, coordinated internationally under the aegis of the United Nations and its specialized agencies and programs. Nongovernmental international organizations of scientists, such as the International Council of Scientific Unions and the International Ocean Institute, also contribute. The marine scientist is the prototype of a new interdisciplinary, international scholar, one who bridges the gap between theory and action. More often than not the marine scientist is an explorer, a sailor, a driver, exposed to physical hardship and adventure, as well as a scholar, a theoretician, a philosopher of nature.

The conquest of ocean space is not merely a feat in itself, an addition to our treasure of knowledge. It transforms this knowledge. It revolutionizes our concepts of the earth, its genesis, its ecology, its resources, man's role on it, and the requirements of its management and governance.

Anaximander of Miletus conceived that
there arose from heated water and earth
either fish or creatures very like fish;
in these man grew, in the form of embryos
retained within until puberty; then at last
the fish-like creatures burst and
men and women who were already able
to nourish themselves stepped forth.

Censorinus,
De die natali

CHAPTER TWO
THE BIOLOGICAL SETTING

IT IS WITHIN THE COLOSSAL GEOPHYSICAL DRAMA OF THE OCEANS, which is also the
drama of the planet earth, that a second drama must be placed—the evolution of life.
Without knowledge of the oceans' changing climes and shifting contours, life's
mysteries would never yield. Only the ocean was capable of giving rise to life and
containing its myriad forms in practically unbroken continuity from its misty beginnings, at least some two billion years ago, to this day.

The conditions for living forms to arise are fairly general and not hard to meet.
For life to begin on another planet, that planet would have to be approximately the
size of the earth. If it were much smaller, it could not retain a suitable atmosphere
around it; if it were much larger, its atmosphere might be so dense as to be
impenetrable to radiation. But considerable leeway exists within these limits.
Temperatures, determined in part by distance from the sun, would have to
approximate those at which water remains liquid. And another, rather broad
condition is that an external source of energy, such as sunlight, must exist, and
radiation must be contained within a wavelength range between about 300 and 1,100
millimicrons (one millimicron is one ten-millionth of a centimeter); shorter
wavelengths destroy macromolecules, while longer ones fail to activate photochemical
reactions. (There might still be nonphotochemical life in such an environment,
however.)

These are the basic requirements, and it seems that there are at least one billion such planets in our galaxy alone. Furthermore, as George Wald has pointed out in his masterly essay "The Origins of Life," there are about 100 million galaxies now within the range of our telescopes. Assuming a somewhat uniform composition of the universe and considering the small number of chemical elements involved, the number of planets suitable for life merely in that part of the universe which we already have under observation would be about 10^{17} (100,000,000,000,000,000).

But let us focus on the planet earth. In its primeval form its atmosphere was made of volcanic gases, poisoned by methane and ammonia. There was no oxygen or carbon dioxide—or anything we associate with the existence of life. The ocean was, as Marston Bates put it, a sort of thin organic soup in which almost anything might happen, and in this mixture of dissolving minerals, large molecular compounds of carbon combined and recombined for millions of years, finally making cells.

One can distinguish four phases or processes in the evolution of cell metabolism and energy production. The first is *fermentation*, whose by-product is carbon dioxide. The second and third phases are somewhat more exotic: the *hexose-monophosphate cycle*, which produces hydrogen and carbon dioxide and is one of the first examples of the metabolic splitting of water in a physiological process that gives the cell the energy it needs; and *photophosphorylation*, which is the direct utilization of sunlight to produce high-energy phosphates, such as adenosine triphosphate. At this point the monocellular being has evolved the pigment chlorophyll with which to trap light, and, to simplify a complex story, the way is now open for the evolution of the fourth process, *photosynthesis*. The cell begins to use sunlight to synthesize glucose, releasing molecular oxygen as a by-product. Thus, oxygen enters the atmosphere, and when it reaches a sufficient concentration, cellular respiration becomes possible. This is what happened about three billion years ago.

Photosynthesis and respiration, then, are inverse or complementary processes. And all the free oxygen in the world passes through organisms—in by respiration, out by photosynthesis—every 2,000 years; all the carbon dioxide passes through organisms in the reverse direction every 300 years; all the waters on earth are decomposed and recomposed by photosynthesis and respiration every two million years.

Cells that "learned" photosynthesis became plants. Cells that did not, became animals that ate the plants. But life remained bound to the ocean forever. Even when the animals left the ocean, they carried with them blood that is like seawater. In most marine invertebrates a solution of ions circulates that is essentially seawater. The ionic composition of vertebrate blood, according to George Wald, so closely resembles seawater diluted three or four times that scientists have assumed it corresponds to

36

seawater in the remote past when our ancestors closed off their circulatory systems.

Life was a great success. It grew mightily and it made its environment as it made itself. Prepaleozoic plankton—especially the blue-green algae that are still with us— filled the atmosphere with enough oxygen to make possible the evolution of the metazoan as early as 10 million years before the Cambrian period, 500 to 600 million years ago.

As their own creation unfolded, living forms created not only the atmosphere but the seascape as well. The skeletons and shells of the tiny, unicellular plankton, both animal and vegetal, eventually sank down to the bottom and lined it with thick sediment.

The next big step, then, was the evolution of the metazoan, the multicellular being, whether plant or animal. This happened in several stages: one was simple "togetherness." Cells divided, but instead of drifting apart, they stuck together, each one feeding itself, metabolizing and reproducing itself, each one an individual, "born equal." Together they formed a colony, for example, a sponge (Porifera).

Perhaps the most curious being, straddling individuality and nonindividuality, is the Portuguese man-of-war; it is definitely a "colonial" animal. The adult man-of-war is in fact made up of several hundred animals. But there is a division of labor— specialization. The subanimals are of four different types. One type—and there is only one of this type for each man-of-war—constitutes the sail; a second type constitutes the fishing tentacles; a third, the digestive tract; a fourth takes care of the reproductive function.

As specialization and organization proceeded, the subindividuals lost individuality. This happened with the emergence of two layers forming a hollow sphere, the endoderm (inner layer) and ectoderm (outer layer), and an oral opening to the central digestive cavity. This is the simple basic structure of coelenterates, such as jellyfish. Next came a middle layer or mesoderm, and the formation of a body cavity (coelomates, worms, in their simplest form, of which the annelid worms are the oldest). The ectoderm developed a nervous system and formed other body structures, at first as an outside armor for soft creatures which really needed no bones to float in the sustaining and supporting waters of the ocean. The mesoderm also took part in this, and internal skeletons emerged when the migration of life away from the oceans had begun (chordates, vertebrates).

The "recorded history" of life starts 600 million years ago with the Cambrian period, from which time well-preserved fossils of hard-shelled animals are quite abundant. But jellyfish, segmented worms, and soft corals go back even further than that. Their soft existence, imprinted in mud and sand, can be documented for periods beyond 700 million years ago. There must have been at least a thousand species in existence in the oceans by then, of which 455 are documented; they include Porifera (sponges), Coelenterata (jellyfish), Echinodermata (cystids and edrioasteroids, sea stars, urchins), Vermes (tubes, trails, and burrows), Brachiopoda, Mollusca, and Arthropoda (trilobites and crustaceans). At the beginning of the Devonian age, 400 to 350 million years ago— while Europe and North Africa were approaching each other and the supercontinent Pangaea was in the making—the fauna of the world ocean was not very different from what it is now.

GEOLOGIC TIME SCALE*

PALEOZOIC ERA	PRECAMBRIAN:	600–? million years ago. Constant earth upheavals. Soft-bodied marine animals.
	CAMBRIAN:	500–600 million years ago. Moderate upheavals. Trilobites, mollusks, first crustaceans. Brachiopods.
	ORDOVICIAN:	425–500 million years ago. Mild climate. Seaweeds, clams, nautiloids, first vertebrates. Early fishes. Life comes ashore.
	SILURIAN:	405–425 million years ago. Mild climate. Ammonites. Many coral reefs.
	DEVONIAN:	345–405 million years ago. Varied climate. Many fish. Amphibians. Insects evolve.
	CARBONIFEROUS:	280–345 million years ago. Warm and humid. Coal swamp forest, gigantic insects.
	PERMIAN:	230–280 million years ago. Extremes of climate. Trilobites become extinct. Reptiles flourish.
MESOZOIC ERA	TRIASSIC:	181–230 million years ago. Arid climate. Dinosaurs, mammal prototypes, first pupal insects.
	JURASSIC:	135–181 million years ago. Mild climate. Giant land and sea dinosaurs, toothed birds.
	CRETACEOUS:	63–135 million years ago. Water inundates much of the earth. Dinosaurs become extinct.

*Adapted from *The Sea,* Life Nature Library, and Nigel Calder, *The Restless Earth*

CENOZOIC ERA	TERTIARY:	1–63 million years ago. Continent building. Modern plants, dominance of mammals.
	QUATERNARY:	0–1 million years ago. Four ice ages. Many mammals become extinct. Man appears.

All the animal phyla, or subkingdoms, that ever existed exist today in the oceans. Some of them are represented *only* in the oceans.

The whole drama of creation, the history of life, runs through the oceans in three dimensions, as it were. There is a diachronic, or through-time, dimension, along which you can plot the evolution of life in the vertical, so to say: beginning at the bottom, with the coming of the first protozoa over two billion years ago, we move through the sponges of the Precambrian; the trilobites, mollusks, and first crustaceans of the Cambrian (500–600 million years ago); the clams, nautiloids, and early vertebrates of the Ordovician (425–500 million years ago); the coral reefs of the Silurian (405–425 million years ago); the fish and amphibians of the Devonian (345–405 million years ago); up to the first appearance of mammal prototypes in the Triassic (181–230 million years ago); the rise and fall of dinosaurs in the Jurassic and Cretaceous (63–181 million years ago); and the conquest of the earth by mammals (and modern plants) in the Tertiary (1–63 million years ago), with man appearing early in the Quaternary, about a million years ago.

There are many points on this diachronic line, many moments in the great drama at which we could stop and enlarge, but we will focus on two as illustrations of the awesome spectacle of the unfolding of life.

One is the origin of the vertebrates, the transition from simple, sessile (not free to move about) filter feeders whose bodies, attached by a stalk to the seabed, consist of nothing but a digestive tract with arms stretched forth to wave drifting food particles into the mouth. There are a variety of such beings, perfecting through the ages the humble art of filter feeding. Eventually the filter-feeding apparatus transformed itself into a set of gill slits whose primary function, it seems, was not breathing (which was done through the skin) but feeding.

At the end of this line of evolution are the tunicates, sea squirts. They were sessile and filter-feeding, and reproduced by budding until some of them took to producing a free-swimming, tadpolelike larva that moved about in search of a suitable environment in which to settle again into sessile, filter-feeding adulthood. The larva had a rather large head equipped with gills, and a muscular swimming tail reinforced by a longitudinal gelatinous chord, the notochord, which was the predecessor of the vertebral column. The motion of the tail was controlled by a dorsal nerve which in the head region received information from rudimentary sense organs. The innovative larval phase was short: after a few hours of free swimming, it would give up and attach itself to the ocean floor. Tail, notochord, and sense organs shriveled into oblivion and were reabsorbed into the adult shape of a tunicate.

Some of the youthful creatures, however, resisted. They became sexually mature, reproduced, and refused, ever more successfully, to metamorphose into sessile adults. Thus, in a very literal way, we owe our backbones to a youth culture.

All this happened in very early Paleozoic times, for in the Ordovician the vertebrates were already well established.

Another, most curious scene in the drama is the transition from fish to reptile. (How, incidentally, did the ancients know that there were fish in the sea before there were beasts on land? Here we have one more example of the strange continuity between myth and science.)

The protagonist in this scene is Latimeria, a crusty-looking fish of the kind called coelacanth, measuring up to 5½ feet and weighing up to 180 pounds. The coelacanths flourished about 400 million years ago, and fossil records pronounced them to have become extinct about sixty million years ago, at the beginning of the Tertiary period. However, they are still with us. A fisherman, casting his net off the coast of South Africa, brought one up in 1938, and quite a number of specimens, about three or four a year, have been caught, dissected, and studied since. The coelacanths are a subgroup of a larger group of fish called crossopterygians, and they form a link between the water and land animals. Coelacanths were the ones that developed legs from their paired fins, and a sort of lung. Being thus equipped for both walking and breathing, they could evolve toward an amphibian existence.

Latimeria, it seems, is no great swimmer; he probably walks or creeps on his fins at the bottom of the sea, among basaltic rock 400 fathoms deep. His eyes are phosphorescent, his color a steely blue-gray that turns to a chocolate brown after death, which comes all too soon after capture, due mostly to the effects of decompression.

There was a second kind of crossopterygian in the Devonian period: the rhipidistian. The rhipidistians are closely related to the first amphibia which arose in the Carboniferous age, 280 to 345 million years ago. Presumably they disappeared during the Permian, about 250 million years ago.

But do we really know? True, the oceans have become a tightly woven net of communications. Nothing, apparently, remains hidden to man's oceanographic ships, submersibles, and unmanned tracking devices, to his underwater photographic and television equipment; nothing remains impenetrable to his sonar and laser or undecipherable to his carbon dating or magnetic measurements. And yet, the oceans are deep and dark. The coelacanth, after all, was rediscovered only yesterday. Perhaps somewhere his brother the rhipidistian is still alive and well. And who knows what else?

The second dimension along which we can plot the history of life in the oceans is synchronic, for all, or almost all, the phyla of life that ever existed still exist in the oceans right now. Thus, we can plot this evolution horizontally, starting from the simplest and moving toward the most complex organism.

The points on the synchronic, or horizontal, coordinate can be ordered according to various principles, and the resulting lines will diverge and converge at various points. We could follow the principle of increasing complexity, from the planktonic protozoan to the marine mammal. Or we could follow the intricacies of the food web, starting from the tiny plants, or diatoms, and flagellates, which in their wondrous variety of shape make up the phytoplankton that drifts about in the lighted surface layers of the open sea. We could move to the tiny animals that feed on this phytoplankton, such as copepods, krill, or crab larvae. These in turn are eaten by larger fish like mackerel or herring, which end up in the stomachs of still larger ones like tuna or swordfish. A sperm whale, it has been estimated, eats a ton of herring, or about 5,000 fish, two or three times a day. Each of the herring will that day have eaten 6,000 to 7,000 small crustaceans, while each of the crustaceans will have eaten up to 130,000 diatoms. Adding it all up, the sperm whale is sustained by trillions of diatoms a day. There may be several million diatoms in one quart (or liter) of productive surface water.

Or we could order these living things in the ocean according to their various habitats: those which live on the sunlit surface of the wide-open seas, such as the tuna, dorado (a dolphin fish), porpoise, or swordfish; those which live in the dimly lighted midwater zone—predators such as the squid, octopus, gulper, or anglerfish, or feeders on the sinking fallout of organic matter, like the lantern fish. Descending into darkness, life becomes sparser, although this sparseness is relative. Living creatures in the deep regions of the ocean make up less than 100-millionth of the water volume surrounding them, 4 percent of the density reached in the most populated regions, yet this still constitutes an enormous total biomass.

The deep ocean floor, finally, is inhabited by a variety of ancient creatures—glass sponges, sea lilies (crinoids), and lamp shells (brachiopods)—reaching as far back in time as down in depth. The most numerous are the sea cucumbers (holothurians).

The third dimension of our complex model might be called microchronic. Along this line we can plot the ontogenetic recapitulation of the history of life in the embryos of the most evolved species. For is not the womb of every woman a micro-ocean, and does not every new life start as a simple protozoan? As a tiny tadpolelike creature—a chordate, cartilaginous animalculum—before turning true vertebrate? As a gill breather before its lungs are activated?

Every point in this rapid development has its corresponding point somewhere in time and somewhere in the wide spaces of the oceans. In the human embryo, for instance, the heart begins as a simple enlargement of the principal blood vessel. The enlarged section divides into four parts, all in a row. The vessel then coils back on itself, and the heart acquires its globular shape, with two sections folded (auricle) over the other two (ventricle). The coelacanth had this kind of linear heart 400 million years ago and it still can be found in the depths of the ocean today.

The human embryo is coelacanthine in yet other ways: for example, the earliest stage of the development of its pituitary gland strikingly resembles the pituitary gland that the coelacanth has retained throughout its half-billion-year life in the oceans.

Such a complex three-dimensional model could of course be built for the whole biosphere, including life in the oceans, where it began; life in fresh water, where it first extended itself when it transcended the boundaries of the sea; and life on land, where it unfolded in such imperious ways. For the biosphere is one and indivisible, and water and air and the mass of continents all circulate. But if we accept the traditional division of the biosphere into water (salt and fresh), land, and air, then the oceans are by far the greatest habitat of life, and the most continuous in time and space. Life that left the ocean, furthermore, has returned and continues to return, like waves that wash the shore and recede. The bony fishes, for example, seem to have evolved in fresh water and returned to the ocean subsequently. Indeed, the only truly indigenous ocean fish are the cartilaginous sharks and rays. The reptiles returned after trying the interface between water and land and leading two lives, *amphi bios*. Birds, having conquered the air, returned, like the penguin. And the mammal, evolved on land, returned: seal, otter, sea lion, dolphin, whale. And man is returning. That is the drama within the drama within the drama.

In the oceans, one gets a different view of life.

One striking element is the incredible number of evolutionary options, whether one looks at the origin of life in general; at the origin of each species; at the origin of social behavior and the origin of sex (or the curious interactions between the two).

Most wrasses, parrot fish, and groupers, for example, change their sex, and the change is controlled by factors of social organization.

Some species of wrasses are known as cleaner fish—the barbers and hygienists of the oceans, who make their living by eating parasites from the skin and gills and inside the mouths of other fish. They live in groups of a dozen or less, consisting of a dominating male and a descending hierarchy of females. When the male disappears, the largest female takes his place—first socially, behaviorally. Then this social and behavioral metamorphosis is followed by physiological metamorphosis, and the female becomes a true male. In fact, all males among the cleaner wrasses are derived from females.

David Guttman, in a recent paper, points to the emergence in our contemporary society of social sex reversal in humans. In view of the wrasses, it is certainly interesting to speculate on the possible biological consequences of this development.

Another striking element in the ocean is the lavishness with which it provides for life's reproduction and continuation. Such is the abundance that if, for example, all codfish eggs that are spawned in the sea matured into adult fish, the Atlantic Ocean would be packed solid with codfish within seven years. A fish spawns millions of eggs at a time, an oyster as many as 500 million a year. The diatom, the so-called meadow grass of the sea, may have a billion descendants in a month. Only a tiny fraction—one-thousandth would be a high figure—survive.

And what is meant by surviving? Surviving long enough to mature? Long enough to reproduce? On the whole, death in the oceans is not a consequence of aging and decay. What we think of as natural death is, paradoxically, unnatural in the oceans. Death means taking one's place in the food chain. It has been calculated that only one out of 10 million marine animals does not end up in the stomach of another.

Fishes fish. And to do so they have developed technologies which are astounding. Our skills are no better than theirs. On the contrary, their natural fishing gear ranges from simple fishing nets to lines and hooks, spears and harpoons, sophisticated chemicals, and sonar for locating their catch.

There is a urochordate, the larvacean, a small planktonic herbivorous creature, low on the food chain and the scale of evolution, ancient in time. It invented the first fishing net. The animal is wormlike, with an enlarged head containing a nervous system, gill slits, a tail, and an anus. The larvacean builds itself a spacious, gelatinous house in which it places a system of large nets made of strong, fine threads. The house has three gates. Two of them serve as water inlet and outlet, maintaining a gentle flow of water through the nets. The third is an emergency exit. When the nets become clogged or otherwise malfunction, the larvacean slips out through the third door and builds itself a new house—with a new system of nets.

The oldest line fisher is the anglerfish. Rod and line, called the illicium, protrude from just above the mouth. The illicium may be four times as long as the animal itself, and is equipped with a light at the end which is used as bait.

Light, incidentally, is used by marine animals not only as bait but for a variety of other purposes: communication, species recognition, locating a mate, or frightening an enemy with strong flashes. The squid *Heteroteuthis,* when attacked, disappears in a fiery cloud of luminescent ink.

How marine animals produce their light is not yet fully understood, but their production system is far more efficient than any man-made one. It engenders no heat waste or infrared or ultraviolet rays—it is light pure and simple, produced by the living cell's excretion of a simple protein called luciferin. The energy released as light is, for biological systems, truly prodigious. How the protein stores all this energy is not completely understood.

The silvery phosphorescence of breaking waves, a common sight on warm summer nights, is caused by countless microorganisms, mostly dinoflagellates of various types. Strange flashing light is emitted by jellyfish, copepods, and other larger forms. A number of fish and crustaceans can voluntarily switch these lights on and off. Six specimens of a Norwegian crustacean in a two-quart jar can produce enough light for a person to read a newspaper, and they can switch it off and on. Some fish, squid, and euphausiids have searchlights with reflectors, lenses, and irises.

In many cases animal technology is far less wasteful than human technology. This is true of locomotive as well as of light-producing energy. A dolphin, for example, can attain a speed of 25 knots. A six-foot dolphin weighing about 300 pounds would need about 14 horsepower to drive it at 25 knots—an output of 87 horsepower per ton, or more than six times the muscle power that a trained athlete can produce. There is in the dolphin's shape, texture, and way of moving a minimal loss of power in kinetic energy.

Right whales, it has only recently been discovered, can sail. They stand on their heads, with their large tails out of the water as a sail. They use their dorsal fins as keel and their flippers as rudder, so they can tack—a technique it took mankind thousands of years to acquire!

But to go back to fishing techniques.

Horrid are the poisons injected with a thousand needles by the sea anemone into a carelessly passing fish. They paralyze, and without further struggle the fish is drawn into the anemone's mouth and swallowed whole. Corals and many jellyfish are also equipped with these stinging cells, which are called nematocysts. They have a long hair which is normally coiled inside them like the inverted finger of a glove. When the cell is disturbed, the barbed hair pops out. Another example of the economy of animal technology: when a sea slug eats one of these animals, it digests the "meat" and saves the still-useful weapon for its own use. The eolid sea slug stores the nematocysts in special sacs on its back, from which it hurls them at its enemies.

Swordfish and narwhal are well-equipped spear fishers. They move rapidly through a school of fish, large or small, slashing vigorously left and right. Then they turn around and, at a more leisurely pace, make a meal of those their spears have wounded.

Besides his keen vision and an extremely refined olfactory sense enabling him to detect a substance in dilutions down to one part in several million, the shark has a "sixth sense" to rely on for locating his prey. This is the so-called lateralis system—common, in more or less developed form, to all fish—which is vibration-sensitive and enables the shark to locate such disturbances as a ship sinking miles away. He can also use it actively to echo-locate objects by the time relations of reflected vibrations he emits himself.

The sperm whale possesses highly developed long-range hunting sonar enabling him to precision-target a squid or fish in the ocean depths from a distance of several miles.

But it is not only the fishing and hunting technology of the beasts in the sea that gives pause. The skills developed in navigating and in communicating are no less wondrous.

No one can really explain how salmon "home"—that is, find their way from the ocean back to shore, to the stream, up over cascades, right to the pool where they were born, there to spawn and, perhaps, die. Experimental work has revealed the extraordinary refinement of their olfactory sense, and it would seem that every salmon brook has its own smell, produced, conceivably, by young salmon that have not yet left the river, in the form of a secretion of pheromones. The olfactory sense, then, would be the main navigational aid of the salmon—as, in general, it is an important means of communication in the ocean. But is it enough of an explanation, considering the vastness of the distances and the multitude of choices? Currents and celestial bodies may also guide the salmon on the high seas.

No one really knows how the green turtles of Brazil direct their navigation to Ascension Island, a dot in the middle of the South Atlantic 1,400 miles away, or why they feel impelled to lay their eggs there, and only there, once every two or three years. Is it because of island smells, carried by ocean currents on which the young turtle travels down and the mature struggles back up? It is hard to prove convincingly.

The turtle must have a compass sense. It may also use the sun by day and the stars by night as navigational aids. We simply do not know. All we know is that the mysterious voyage does take place: thousands of turtles have been tagged and many have been recovered. Tracking tests have been made using turtles to tow floats from which helium-filled balloons rose to mark the position of the migrant. But only short distances could be monitored in this way. More recently, radio transmitters have been mounted on the turtles' backs, and NASA satellites will soon be used to track their signals over great distances. Thus, the most sophisticated human technology must be employed in the attempt to understand the natural technology of this ancient animal, a technology which otherwise eludes our comprehension.

The eels of the Atlantic—so it would appear—have an international organization, with an annual convention in the Sargasso Sea near Bermuda. They arrive there all the way from the rivers of Europe as well as from the Americas. They dive and spawn there, in the thick floating weeds of this current-bounded ocean within an ocean—itself a mystery to science and fraught with lore and legend, a terrible trap where plant filaments and seaweed grip vessels in an unbreakable net. Then the young larvae depart, and though they all seem alike, those of European parentage take up the long journey, riding the Gulf Stream and the North Atlantic Current for three years to the coast of Europe. The American members of this puzzling organization have it easier. It takes them only six months along the Gulf Stream to make their way into the American rivers from which their parents came. The larvae have grown into elvers by the time they reach their destined rivers, in which they take up their freshwater life, eternal waves of oceanic life invading the freshwater ecosystem, to be washed back again into the ocean. After ten years they embark on the long, hazardous return voyage to the faraway Sargasso Sea. What signal starts them, what compass guides them, no one knows.

Human achievements shrink considerably in impressiveness and significance when one studies the ocean environment and the works wrought by the animals of the sea. Even the changes in the environment introduced by man, of which we are so painfully aware today, are puny in comparison with those effected by the marine fauna. Have not the fauna created the very atmosphere which we are just beginning to tamper with? Have they not built mountains in the billions of years of their existence and covered the ocean floor with their myriad remains?

When we play god, hurling atomic fire, shaking the fundaments of the earth and threatening an end to all its living forms, we should pause instead, and wonder, and look once more into the mirror of the ocean. Life there is tough, much tougher than one might think. Time and again it has survived ecocatastrophes as great as any we could produce, and started all over again.

Take as an example the history of the coral reef, the oldest ecosystem on earth, which had its beginnings over two billion years ago.

Drilling into the depths of time through the masses of matter built up layer upon layer and reading the fossil record imprinted into them is like excavating an ancient city and reconstructing its history through the centuries: conquests and incursions of enemies, triumphs and epidemics, and the way of life of its inhabitants in various epochs. Only the time scale is different.

The reefs' flora has been constituted through the eons by various types of lime-secreting algae, blue-green, green, and red. Their stony deposits intertwined with the coral skeletons which formed the core of the many layers, from the dimly lit bottom caves up to the surface. The whole structure was cemented by the fine detritus of tiny plants and animals.

The oldest reefs are the simplest, built by algae alone. They go back billions of years to Precambrian times, and can be found in rock formations in all parts of the world. Called stromatolites, their blue-green algae were closely related to those still surviving today. Over millions of years, individual colonies grew upward for tens of feet.

In the early Cambrian, over 600 million years ago, the first animals associated themselves with the reef. Cuplike stony beasts called archaeocyathids emerged like low trees from the meadows of algae. Bottom-feeding trilobites crawled around in between.

This form of community life did not last long, however. By the end of the Middle Cambrian, about 540 million years ago, the archaeocyathids were wiped out by some as yet unexplained ecocatastrophe. There were no reefs for about 60 million years. But what are 60 million years? Bigger and better reef communities made their appearance in mid-Ordovician times, about 480 million years ago.

Stromatolites still formed the basis, but coralline red algae and also the first sponges—shaped like encrusted plates, or hemispheres, or shrubs—joined the community. And the first real corals emerged.

Near the end of the Devonian period, just over 350 million years ago, worldwide climatic changes decimated life in the oceans. In the coral reefs, only the stromatolites survived for the next 13 million years.

The Carboniferous age, followed by the Permian, brought the reign of the bryozoans and brachiopods over the reef community. Chambered, sphinctozoan sponges and sea lilies began to play an increasingly important role.

All this was wiped out by a new, even worse ecocatastrophe at the end of the Paleozoic, about 230 million years ago, and there were no more reef-building communities for another 10 million years. Then, in the mild climes of what is now the Mediterranean a new reef community emerged: in mid-Triassic times the scleractinians, the progenitors of the more than twenty families of corals living today, made their first appearance in Germany, the southern Alps, Corsica, and Sicily. Sponges, sea urchins, mollusks joined in and flourished. But about 100 million years ago a revolution took place. Its protagonists were the rudists, a group of bivalves with cylindrical and conical shells that built upward as though in imitation of coral-building patterns. Starting from nowhere, the rudists cemented their power over the next 60 million years. They dwarfed the corals and repressed their building. Then, some 65 to 60 million years ago, they disappeared as suddenly as they had come.

This, again, was a period of great dying in the oceans. Nearly one-third of the animal families that flourished in the late Cretaceous period no longer existed at the beginning of the Cenozoic. Many corals, mollusks, clams, and sponges went down with the rudists at that time. Temperature changes, the drying up of shallow seas, the emergence of harsh, seasonal, continental climates may account for it, together with factors still unknown.

From their cosmopolitan ubiquity of earlier ages, reef-building communities were now more restricted, essentially to areas between 35 degrees north latitude and 32 south. The richest developments now are in the western Pacific, the Indian Ocean, and the Caribbean. The most stupendous creation is the Great Barrier Reef off Australia.

The modern reef community consists of algae, corals, sea fans, sponges, sea anemones, tunicates, barnacles, mollusks, and the like, and a great variety of fish. Butterfly and anglerfish, damsel and squirrelfish, parrot fish, triggerfish, wrasse, grouper, and grunt, pipefish, snake eels, and scorpion fish all find food, shelter, and oxygen in the reef caverns and branches—an exuberance of colors and shapes that is matchless in nature.

Population eruptions of starfish recently threatened the Great Barrier Reef of Australia and other reef regions in the Pacific. The phenomenal increase in starfish was thought by some to be due to manifold human influences. Others ascribed it to natural causes. Only the time perspective of our drama will reveal whether one or both causes were involved. (It seems, at any rate, that the worst is over, at least for the time being, and the reefs are regenerating.)

Eutrophication is killing the reef's myriad forms of life on Oahu, Hawaii; in the Caribbean they are affected by oil. The corals die. The tiny polyps that make up these vast colonies cease to extend their tentacles to filter food from the waters' microscopic life, and the tiny green algae living on the surface of their bodies die. Fungi take their place, or else more robust algae cover and suffocate the corals. The scene becomes dark and dull. The fish move away or die, and erosion takes its course.

This time we know the cause of the disaster. The rudist of the Quaternary is man, the rudest. But he, too, will go as he has come. The conclusion of the best experts on the history of reef communities is: "Any collapse of the present reef community will surely be followed by an eventual recovery. The oldest and most durable of the earth's ecosystems cannot easily be extirpated" (Norman D. Newell, "The Evolution of Reefs"). It may take millions of years for the reef community to overcome the ecocatastrophe man has wrought. But what are millions of years?

48

Methought I saw a thousand fearful wrecks,
Ten thousand men, that fishes gnaw'd upon;
Wedges of gold, great anchors, heaps of pearl,
Inestimable stones, unvalued jewels,
All scatter'd in the bottom of the sea.

William Shakespeare,
King Richard III

CHAPTER THREE

THE MAN-MADE SETTING

STAGE AND DRAMA ARE INTERACTING. The physical setting of ocean and earth is involved in its own drama, into which the second drama, that of biological evolution, is being woven, constantly remaking the physical background.

The drama of mankind is just a minute episode in the drama of life within the drama of ocean and earth. No matter how small, however, the human drama feeds back on the others. Man remakes himself as he remakes his environment.

Pieces of the human drama have become part of the physical setting of ocean and earth. Seabeds and submerged coastal areas are strewn with the remains of human industry, ingenuity, and courage. The hulks of sea-going vessels—literally tens of thousands of them—litter the seabeds. Galleons rammed and sunk in ancient battles, sailboats defeated by winds and waves or run afoul on uncharted reefs and rocks.

The harder they are to find, the better they are preserved. Wrecks lying open on the seabed are likely to disintegrate in a relatively short time, their wooden beams and hulls eaten by worms, their cargoes rotted and dispersed. Wrecks buried in mud and silt, however, have resisted millenniums. Sheltered from the inclemencies of nature and the depredations of man, they have had a far better chance than have the terrestrial remains of ancient civilizations.

Ancient cities and harbors have been engulfed by tidal waves in the wake of earthquakes. Others sank through the centuries as earth and ocean heaved and subsided and water levels rose and fell. Local fishermen gazed on sunken cities through the deep-blue Mediterranean waters which enlarged their ancient contours and made their stones seem to quiver—until they disappeared and became legend.

Our descendants may some day glide over Venice as we now glide through its canals, and marvel at its silent palaces and the fish fluttering around St. Mark's the way pigeons do today. Then, during what may turn out to be the second Dark Ages, mud and sand and silt, reef builders and encrusters, wastes and deposits will do their work and erase the memory of one of the greatest monuments of human civilization. But then, perhaps, the archaeologists of a new era, lured by legends passed on by local fishermen, will arrive on the scene with aqualungs or in submersibles. They will establish their sea labs and domes down there, pulse the seabed with metal-sensitive magnetic or acoustic devices, excavate, uncrust, photograph in three dimensions, and map and measure. They will raise the city from the water in bits and pieces with air lifts or lifting balloons, apply chemical treatments to keep it from warping and shrinking, and then redesign and recompose. It will be a herculean labor, multi-faceted, interdisciplinary, calling on the skills of sportsman, explorer, archaeologist, physicist, engineer, chemist, physiologist, photographer, designer, historian, and linguistics specialist, among others.

And the results will be rewarding. It may be too much to say that marine archaeology has revolutionized our concepts of ancient history and prehistory the way marine geology has revolutionized our concept of the earth; or perhaps it is too early to say this. Marine archaeology is a young science, hardly come of age, and no one can tell what it has yet to reveal. But it has already shown us a great deal we would not otherwise have known about the amazing technological achievements in naval and port design and construction of very early periods of history; the existence of astronomical computers as early as the first century B.C. or even earlier; trade routes and cargoes; life styles aboard ship; art and custom.

The prime sites of exploration thus far have been the Mediterranean, the Caribbean, the Baltic, and the Florida Straits. The geographic limitation of these sites leaves ample room for speculation as to what may still lie hidden in the rest of the ocean world.

Among sunken harbors, the most famous are the Phoenician ports of Sidon and Tyre at the easternmost point of the Mediterranean; the Roman port of Caesarea on the coast of Israel; the Greek harbor of Apollonia on the Libyan coast; Chersonesos in Crete; and Cherchel in Algeria.

Sidon and Tyre were explored by a French Jesuit, Père Poidebard, as early as 1935–37, well before the invention of the aqualung, with very limited technological means. George Bass has described this work in his book *Archaeology Under Water:*

50

Lava flowing into the sea off Hawaii

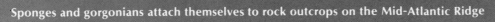
Sponges and gorgonians attach themselves to rock outcrops on the Mid-Atlantic Ridge

The submersible *Deepstar 4000* being lowered into the sea
off Cape San Lucas, Lower California

Hi Kino, a deep submersible made of glass, being moved to its site

Fungia and coral fluorescing

Great Barrier Reef, near Lizard Island

Emperor angelfish (*Pomacanthus imperator*)

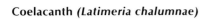
Coelacanth *(Latimeria chalumnae)*

Tube worm *(Sabellastarte magnifica)*

Polkadot grouper *(Cromileptes altivelis juv.)*

Mandarin fish *(Synchiropus splendidus)*

Portuguese man-of-war *(Physalia physalis)*

Masked booby *(Sula doctulatra)* with young

Onespot fringehead *(Neoclinus uninotatus)*

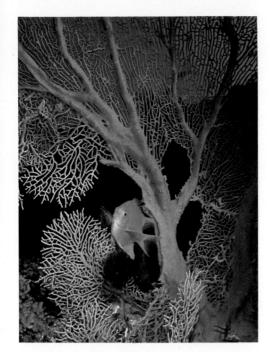

Sea fan and damselfish

Marine iguana *(Amblyrhynchus cristatus)*

Porpoises surf-fishing off Lower California

Nudibranch *(Hermissenda crassiconnis)*

Zoanthids *(Parazoanthus swiftii)* living on a black sponge

Tube-building fan worm *(Sabella melanostigma)*

Overleaf: Magnificent sea urchin *(Astropyga magnifica)*

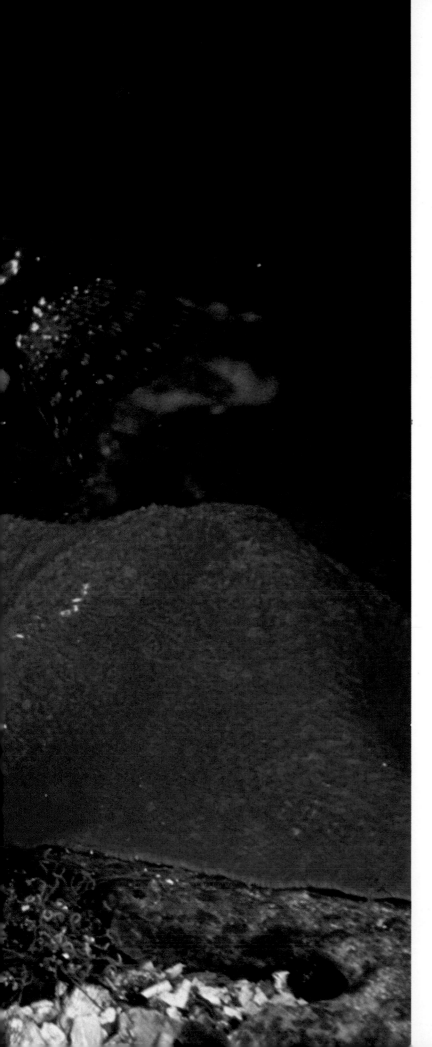

◄ Two bat starfish *(Patiria miniata)* discover each other

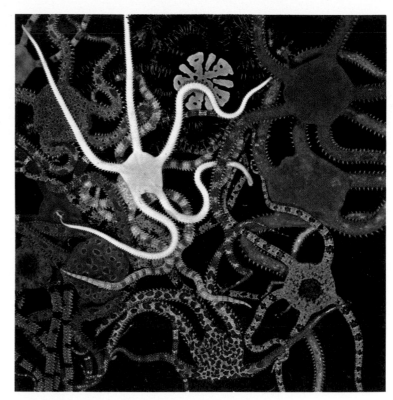

Brittle stars *(Ophiothrix, Ophioderma,* and others)

Brittle star *(Ophiothrix suensonii)*

Bioluminescent copepods

Jellyfish (*Pelagia noctiluca*)

Purple jellyfish *(Pelagia panopyra)*

California halibut larva *(Paralichthys californicus)*

Golden sea horse *(Hippocampus kuda)*

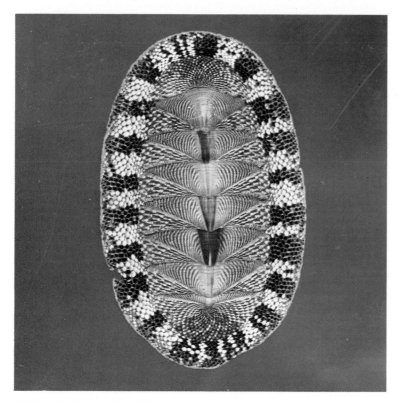

Chiton *(Chiton tuberculatus)*

Slipper lobster *(Scyllarus* sp.)

Fire sponge *(Tedania ignis)*

Inspecting an unexcavated cargo of amphoras off the island of Rhodes

Raising a statue in the Blue Grotto, Capri

Building foundations seen through the water at Kenchreai

The Temple of Isis at Kenchreai, exposed from the sea by the use of sandbags and a pump

The gun carriages of the *Wasa*

"Aerial photographs revealed the remains in shallow water. Helmeted divers, sometimes guided by a local skin diver holding his breath, measured moles and took underwater photographs—the first made for archaeology—of masonry. The shallow depth allowed the placing of buoys on pertinent points, and these were plotted from the surface by ordinary surveying methods. Remarkably, Poidebard took stereo-photographs through a glass-bottomed bucket held on the surface, and with these he was able to study the walls in three dimensions."

The results were impressive. As another scholar-diver, Philippe Diolé, commented: "We know now what the size of a Roman colonial harbor in the second century A.D. really was. We are in a position to see what huge concrete constructions were used to protect the anchorage from incoming rollers, often at some distance from the land; how communication was established between a complex system of basins so designed as to accord with the traditional siting of a place which native experience had submitted to prolonged tests: different channels suited to different winds, the setting of store-houses, tanks and arsenals, the positioning of harbor-craft at the different quays."

The work at Sidon also revealed an arrangement of sluice gates and flushes to prevent silting. This was first done by the Phoenicians, probably as early as the twelfth century B.C., and then taken over by the Romans.

The ancient Greek city of Helike drowned in the gulf of Corinth in 373 B.C.: "First the ground was shaken to the depths by an earthquake," the Greek historian Pausanias reported in the second century A.D. "Then suddenly it opened up and everything which was built upon it collapsed and plunged into the depths, no trace remaining thereafter. Thus perished Helike. It is said that this earthquake was followed by another disaster, caused this time by the season of the year: the overflowing of the sea, which inundated the town and the surrounding countryside. The sacred wood of Poseidon was so submerged that one could hardly see the tops of the trees therein. The wrath of the gods struck the unhappy town through two elements, first shaking it to the ground and then engulfing it with all its inhabitants."

For centuries the ruins of Helike remained visible through the clear water of the Mediterranean, and a number of Greek and Roman writers recall the vision of temples and columns beckoning from the bottom of the sea.

Silt and hard mud now cover all traces of the ancient site, which will probably yield a rich reward to excavators in the future: a whole town dating from the fourth century B.C., with its ramparts, the furniture of its homes, the statues in the sanctuaries, and the skeletons of its inhabitants.

Port Royal, the pirate city on the island of Jamaica, which was sunk by an earthquake on June 7, 1692, was in no better shape when it was discovered. Its buildings were buried in mud under twenty to forty feet of water, leaving no trace even of its walls and towers. Yet Port Royal was rediscovered by two enterprising marine explorers, Edwin and Maria Link, in their specially designed research ship *Sea Diver*. Relying on echo-sounding techniques, they produced a complete map of the old city and proceeded to retrieve household and art objects from the sites where they were most likely to be found. Thus, pottery, brass, and wrought iron wares were brought to light—even a clock, encrusted past redemption, which under X-ray revealed the exact hour and minute at which it was stopped by the earthquake.

The most daring exploits, the most arduous work, if unaided by the sophisticated technologies of our age, could bring to light only one ship in a decade, one epoch-making archaeological discovery in a generation. Only so much could be plotted down there during one submersion; only so much could be raised.

   With the newest types of submersibles, air lifts, pressure lifts, drill pipes, and pick-up tongs at our disposal, work has been accelerated to an unbelievable degree. What would have taken a dozen divers an entire season to explore can now be plotted in a single hour by two experts sitting comfortably in a research vessel.

   One such research ship is the *Asherah*. Named after a Phoenician goddess of the sea, Bass tells us that this ship was "designed to carry a pilot and an observer safely to depths of 600 feet. These men sit inside the vessel's pressure hull, a five-foot sphere pierced with six viewing ports. Surrounding them are such instruments as indicators for speed, depth, air pressure, oxygen and carbon dioxide, as well as a volt meter, gyro-compass, sub-to-diver communication system, sub-to-surface communication system, and a panel of electric switches for interior and exterior lights and cameras. A ton of electric batteries, mounted along with ballast tanks and compressed-air tanks in the submarine's conical tail, can power the side-mounted motors for up to ten hours."

   Objects weighing as much as 200 tons can be lifted from depths of 6,000 feet, and smaller objects, up to 50 tons, from as deep as 18,000 feet, by ships like the *Alcoa Seaprobe* and the *Glomar Explorer*.

As for sunken ships, their number—like that of the dead—is legion. Were we to raise them all, they would make up a fleet more impressive than any existing today, if the few that have been brought up so far are any measure.

   The oldest ship ever recovered is a Bronze Age ship dating from about 1200 B.C. The small sailing vessel—about thirty-five feet long—was discovered in 1960 by Peter Throckmorton at Cape Gelidonya in the Aegean Sea, under ninety feet of water. She had run aground, her bottom ripped by a huge boulder. The courage, patience,

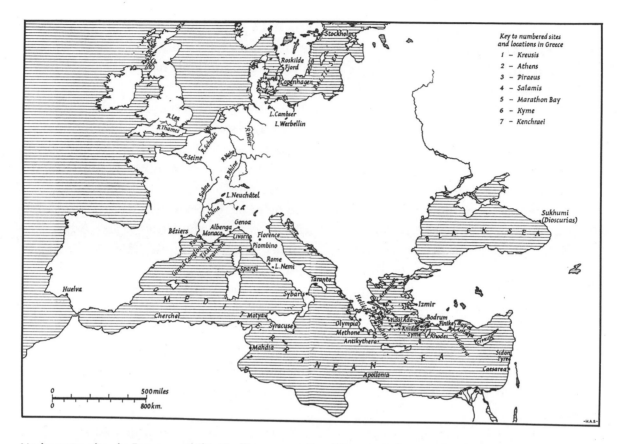

Underwater sites in Europe and the Mediterranean. From George F. Bass, *Archeology Under Water*

and ingenuity it took to raise, uncrust, and reassemble her are beyond imagination. The work was directed by George Bass, assisted by Throckmorton, Frederic Dumas, and Joan du Plat Taylor. Eight divers, holding on to rocks with their legs to keep the strong underwater currents from sweeping them away, worked on the site for weeks with chisel and hammer to detach the wreck from the seabed where it had settled. The detached lumps—up to 400 pounds of them at a time—were raised to the surface by air-inflated plastic balloons. The crusts were peeled off on shore and the parts reassembled, including even some wood which had been preserved.

At one end of the ship there had been a cabin, which contained personal effects such as a razor, an oil lamp, and marvelously precise balance weights. The cargo carried by the ship weighed more than a ton—the largest supply of preclassical copper and bronze implements ever found in the Aegean area.

The analysis and interpretation of this mass of material, by the archaeologists on the spot in cooperation with the London University Institute of Archaeology, was also an astounding feat, a detective story unraveling the tale of a merchant/factory ship which, once upon a time, sailed from Syria to Cyprus. There it took on a cargo of scrap metal, and proceeded westward. The crew must have worked the metal en route, for various tools were found, as well as the two ingredients needed to make new bronze—ingots of copper and of tin. These are the earliest ever found.

The ship must have been Canaanite or Phoenician, which somewhat disconcerted historians, who had dated such Mediterranean voyages somewhat later in the Homeric period, at about the eighth century B.C. But here it was, a ship much like the one on which Odysseus sailed. A few dozen finds like this, and one could relive the millennial history of the Mediterranean, that cradle of civilizations.

Much has been done. Much more is left to be done. Greek ships, Roman ships, Byzantine ships have been discovered at Antikythera, near Marseilles; Yassy Ada (Turkey); Mahdia (Tunisia); and Huelva (Spain). Marble columns, bases, and capitals; mosaics; amphorae galore; statues and statuettes; swords and spearheads; arrowheads and fibulae—there is much more antique art buried under the blue waves of the Mediterranean than exists in all of the world's museums.

Treasures sunk with Spanish galleons on the Spanish Main have been raised off Bermuda. Here is a glowing description by marine archaeologist Mendel Peterson ("Underwater Archaeology" in *Exploring the Ocean World—A History of Oceanography*):

> Here, in a private home in Paget, we saw the great treasure of Spanish gold and jewels: a forty-ounce bar of gold; two round gold ingots, four inches across and half an inch thick; two cubelike sections of gold bar; several single and triple pearl buttons; and, dominating this impressive and glittering array, the emerald cross—breathtaking in the fine color photograph I had seen in New York, unbelievable in actuality. The finely wrought cross, some four and a half inches high, was studded with seven fiery emeralds—symbolizing the seven last words of Christ—of the finest color and luster. . . . At the time I did not know the monetary value of the cross, but I did realize that it was great enough to make it the most valuable single object to have come from the sea in modern times.
>
> After drinking in this overwhelming sight for an hour or so and hearing Tucker's fascinating tale of how the treasure was found, we went across the harbor to a storage building and there saw some of the artifacts from the vessel: pewter, remains of matchlock muskets and arquebuses, a breastplate, black palm-wood bows, and a beautifully carved staff of office for an Indian chief. . . . Navigators' compasses and brass-enclosed time glasses were impressive testimony to the skill of the instrument makers of the time. . . . The collection taken as a whole is one of the most important of the period to have come from any site, wet or dry. The fact that it was from a single ship added tremendously to its

value to historians, for here was a neat slice of late sixteenth-century maritime life, history in three dimensions, dropped into the sand and preserved for modern historians to study.

The Baltic Sea presents a special challenge. Because of its peculiar hydrology—salt water flowing in at the bottom, less saline water flowing out on top—the water is divided into two layers, separated by what is called the "halocline." Life exists almost exclusively in the upper layer, which is more richly oxygenized. The lower stratum lacks oxygen, and with man-made pollution aggravating an already precarious situation, it may end up totally disoxygenized. Unless drastic steps are taken, the Baltic Sea may soon be a dead sea. What is bad for the living, however, may be good for the dead. Low saline content even at the bottom, very cold water, and lack of oxygen prevent the teredo worm and fungus from living and thus help ancient wrecks, which are normally attacked and destroyed by these shipworms.

True, the Baltic is a relatively quiet sea, not so battered by storms as the North Sea or the great oceans. Yet how many ships, fogbound, must have run afoul, how many sunk in battle? The ancient Russians' yearning for sea-lanes and ports; the Vikings' exploits; the Baltic merchants' to-and-fro; the inroads of pirates; the rise and fall of the Hanseatic League; the changing fortunes of the Danish fleet—all must have left bits of life embalmed in the protective deeps of the Baltic Sea.

Some sites of early ships are already known, and will be explored in the future. A few spectacular recoveries have already been made. Most famous, perhaps, is the raising of the *Wasa*, a Swedish warship that sank near Stockholm harbor on her maiden voyage in 1628. The *Wasa* was a 180-foot vessel that displaced 1300 tons and carried sixty-four cannon. In 1664 a Swedish diver, Hans Albrekt von Treileben, descended in a bell to visit the ship and raised about fifty of the heavy cannon from a depth of over 100 feet. Then the *Wasa* was forgotten. Rediscovered in 1956 by Andres Franzén, a petroleum engineer for the Swedish Navy, the ship was finally raised and floated and towed into Stockholm harbor in 1961—a triumph of engineering, physics, chemistry, and conservation. A museum has been built around the *Wasa*, and careful analysis of the thousands of objects found in her is going on apace.

"The *Wasa* is a unique find of the greatest historical importance," Mendel Peterson comments, "and it provides a perfect example of the value of archaeology in providing data that documents alone cannot. Here we have a complete package of maritime history in remarkable condition and containing everything that went into the maintenance and operation of a large warship of the early seventeenth century. The study of the ship and the collections from her will answer many questions that have been asked by maritime historians."

Marine archaeology combines the inquisitiveness of the scientist with the combativeness of the sportsman, the challenge of history with the challenge of nature. If the land archaeologist is characterized by patience, the marine archaeologist is characterized by doggedness. Doggedness in defying a hostile environment and the destructiveness of the elements. Sometimes such doggedness creates an imbalance between input and output, effort and result. Sometimes a mountain of risk and toil delivers something of a mouse in terms of archaeological significance. But at other times doggedness succeeds where patience, tempered by reason, would falter. Feats like the salvage and reconstruction of the wrecks off Cape Gelidonya or Yassy Ada are practically unparalleled on land.

And there is more to be discovered now under the sea than on land. There is more of everything in the sea, and man's return to the sea is just at its beginning. Technology, which is part of human evolution, is about to make man, if he wishes, an aquatic animal, a marine mammal. Captain Jacques Cousteau predicts that human lungs, filled with an incompressible fluid, can be inactivated; blood can then be circulated through a regenerating capsule or artificial "gill" by the diver, who will thus be enabled to descend to any depth, for any duration of time.

According to at least one respected scientist, Sir Alister Hardy, this will be just one more episode of Homo sapiens' changing relationship with the sea. The human species is an odd one, not really knowing its place in this biosphere, oscillating between social animal and loner, carnivore and herbivore, marine mammal and terrestrial animal, bird, beast, and angel.

It is intriguing to consider, as some scientists have, that man, the "naked ape," may have been an ocean dweller in warmer seas. His very nakedness hints at this possibility, for whereas land animals are hairy, marine animals are sleek. Man may have evolved his upright posture supported by the water, a far easier task there than on land, where it would have been a quite unnatural strain. The facility and naturalness with which Homo sapiens swims, the subcutaneous layer of fat he tends to accumulate, in contrast to most terrestrial mammals—all this seems to corroborate the theory that human beings swam before they walked. And is not mythology full of creatures combining fishtails with human torsos and heads? All of mankind may be like Hans Christian Andersen's little mermaid, who paid a high price in suffering for her love for terra firma.

Man's technological ability to "dive into the past" and dwell in its deep recesses must greatly increase if he is really to get to the bottom of it all, a bottom extending over 70 percent of this earth and hiding, perhaps, more traces of man's history and prehistory than we can fathom today.

PART TWO

THE ACTORS

Flounder, flounder in the sea
Prythee hearken unto me:
My wife, Ilsebil
Must have her will
And sends me to get a boon of thee.

After Grimm, "The Fisherman
and His Wife"

CHAPTER FOUR
THE FISHERMEN

THERE IS AN OLD STORY FROM THE FRISIAN ISLANDS, retold by Grimm in "The Fisherman and His Wife." Its theme, which appears in the folklore of many peoples, sums up the story of all fishermen through the ages, from their simple beginnings up to the present point of man's technological clash with nature.

A poor fisherman, so the story goes, is sitting on a rock one day when he catches a large flounder on his line. The fish, amazingly enough, begins to speak with human voice. Claiming to be an enchanted prince, the fish begs the old man to spare his life and let him return to the depths. The fisherman agrees and releases him unharmed. The fisherman's wife, when told the story, is angered that her husband did not ask the enchanted fish to grant them a boon. She plagues the fisherman to find the flounder again and to demand a neat new cottage to replace the hovel in which they live.

Next day the fisherman calls out and the fish swims up. The old man makes his wish, and when he returns home he finds a neat new cottage where the hovel had stood. But the wife is soon dissatisfied. She urges her husband to ask the flounder for more: first for a castle and vast estates, then to become king, then emperor, then pope. The flounder fulfills every wish. But finally she demands power over the sun and moon, to control the universe. This is too much. Amid tremendous waves and crashing thunder, the enchanted flounder destroys all he has bestowed. Empire and kingdom disappear; the fisherman and his wife end as they began, sitting in their hovel in poverty.

The history of mankind's fishing experience parallels the folk story. The poor fisherman—and most real ones have been as poor as the one in the fable—was probably well content with the fulfillment of his first modest wish for better snares. He appreciated his new netting and line, his improved trawls, traps, and seines. But as time went on, the technological imperative pulled him inexorably on to expand his fisheries to distant waters all over the world. He became king and emperor of motorized fleets spanning the world oceans. His arsenal included planes and earth satellites with infrared scanners to map the water temperatures. There were cameras to spot schools of fish, so that he could hunt not only on the surface and the seabed, but could also sweep the midwater with teleguided nets. There were great factories that stayed at sea for months on end processing the fish.

And soon no effort was involved, nor much luck needed, as he wished for and received catches heretofore unimaginable, fishing indiscriminately and ruthlessly for anchovy and whale alike.

But, like the poor fisherman in the fairy tale, man was not destined to control sun and moon or to become Lord of the Universe. That final wish was hubris, begotten of good fortune and begetting misfortune. Man had by this time so damaged and depleted life in the oceans by overfishing and polluting that the fish grew fewer and fewer, and were foul when caught.

The fisherman and his wife were, in truth, poor once again.

In his first role in the ocean, Homo sapiens was probably a fisher. Swimming in the warm sea with the playful dolphins, he was fairly high up in the food chain, eating and being eaten. A spearfisher, probably, like the swordfish and the narwhal. Or a diver for abalone and sea urchin. He may have used obsidian flints to spear the fish or to pry shells open. Like the otter, he may have swum on his back while cracking sea urchins open against a stone. The marine environment, it seems, encourages the using of tools.

All the other technologies—hook and line, net, harpoon, sonar—that the creatures of the deep had perfected over millions of years, man had to reinvent from his land base.

Recent research by nautical archaeologists shows that the earliest cave dwellers in the Mediterranean region in Mesolithic times were hunters feeding on red deer. By the seventh millennium they had become fishermen, as indicated by the large number of fish bones found in their caves. Three thousand years before our own time, fishing had developed into a highly organized craft. Miniatures discovered in Minoan houses destroyed in an earthquake about 1500 B.C. show boats full of fishing tackle, rods, and hooks, and divers plunging into the sea with their bags, one of them carrying what looks like a large sponge.

Bronze Age Mediterraneans probably fished with gorge hooks—that is, bones attached in the middle to a line and sharpened at both ends. The bait was wrapped around the bone, and when a fish swallowed it, the hook was jerked into position in the fish's gullet. Hooks of this type have been found in Mesolithic and Neolithic caves. They were used by American Indians until modern times.

The Chumash and Kwakiutl Indians of the Pacific coast of North America, whom European explorers came upon in the sixteenth century, were extremely primitive fisher folk not unlike those who lived around the Mediterranean 10,000 to 20,000 years ago. The sea was bountiful. Six species of abalone crowded the kelp-enshrouded rocks, and when the tide went out, the Chumash could rise from a shady resting spot and pry a few dozen of the tasty creatures from their precarious moorings. They also made hooks, which they baited with squid and weighted with stones to sink them to where the halibut moved, a few feet off the bottom of the sea. Perhaps they brought the craft of line and hook making with them when they crossed the Bering Strait some 20,000 years ago or sailed the Pacific on vessels like Thor Heyerdahl's *Kon-Tiki*.

The Nordic people were fishermen in the Stone Age, 10,000 years ago, venturing far out in their skin boats after deepwater fish such as cod, pollack, and ling.

Through the ages, the Eskimos, Polynesians, and Melanesians used hooks and lines, and they were sophisticated enough to use artificial baits. The Eskimos carved and painted colorful ones that resembled beetles. The Polynesians used shining mother-of-pearl; did they know of the anglerfish's luring light at the end of its inborn hooked fishing line?

Net fishing probably evolved similarly in the East and the West, with an observant spearfisher noting that the quick fish that darted away at the moment his spear was released could in the end be trapped if he could seal off some area with woven reeds or grasses. The craft of weaving grasses into traps later grew into true net making when man the cultivator developed various materials for making line, such as cotton and wool. With the aid of fibrous plants, especially the one from which common hemp is made, he became a more efficient fisher. Did he know about the larvacean and its net?

It was a big step from spear- or line-fishing to net fishing. The fisherman now took more fish than he and his immediate family could consume. Net fishing brought him to the market and made him a trader. But although traditionally the trader is wealthy, the fisherman remained poor—he is poor in the lore of Nordic ballads, in Biblical parables and Indian tales, in the *Odyssey* and *The Thousand and One Nights*.

Net fishing, furthermore, is a communal or community-orienting activity. Toolmaking and mending, fish hauling and processing are tasks in which the whole village, young and old, male and female, can share. Those going to sea must act as a well-disciplined unit in braving the challenges and dangers of the elements. Those staying behind are united in their anxiety for loved ones out in wind and wave, united in their grief when disaster strikes.

The apparently inexhaustible nature of the fish supply, together with the necessarily collective methods of the fishing enterprise, discouraged the formation of private or individual property concepts. But they did not discourage the technological imperative that drove our fisherman in search of better materials and implements—trawls, traps, lines, and seines.

Unhappily, technological change often does more harm than good. This applies even to very simple changes, such as the introduction of nylon nets. In Brazilian fishing villages, for example, this kind of innovation made the formerly free fishermen dependent on the importer and his middlemen, led to the depletion of the fish supply, introduced competition instead of cooperation, and disrupted communities. This is well documented in a study by John C. Cordell. "We should not delude ourselves," Cordell warns, "about the miracles technology can work in primitive settings. Since modernization is a socially selective process, the question remains: Who benefits from it, and at whose expense? In a highly stratified society like northeastern Brazil, the answer to this question is a foregone conclusion, since part of the cost of technological change is the maintenance of social inequality."

Regardless of its social impact, the new technology allowed fishermen in various countries to stop fishing indiscriminately and, instead, to specialize more system-atically in certain varieties—cod, haddock, pollack, whiting, and hake; on flatfish such as sole, flounder, halibut, plaice; on salmon, sea bass, and rockfish; on fish that travel in schools, such as mackerel, tuna, herring, sardine, anchovy, menhaden, and shad. And the greatest challenge of all, making fishermen drunk with visions of wealth, like dreams of a gold rush or an oil strike, was hunting the giant whale.

Whales had been hunted for many centuries before the great whaling ships were built late in the nineteenth century. In Europe, large-scale whaling began in the Bay of Biscay. The Basque whale fishery was well established in the twelfth century, and is probably at least two centuries older than that.

In 1150 King Sancho VI of Navarre, called "the Wise," made a grant to the city of San Sebastian which listed articles of commerce and the duties to be paid for warehousing them. This list includes whalebone (baleen) as a prominent item. The Basques were expert whale hunters, navigating their coastal waters in small boats and using hand harpoons; they towed the whales they killed to shore for processing. By the sixteenth century, an average of at least sixty whales a year were taken along the Basque coast. This seems to have been a case of overfishing, for the stock began to decline in the seventeenth century. The Basques then sailed to remote places such as

Newfoundland, in what seems to be the first example of distant-water fishing. British, Russian, Dutch, Spanish, and German ships soon followed, and by the middle of the seventeenth century whaling in Spitzbergen and around Bear Island and Jan Mayen Island was big business.

From the Arctic to the Antarctic, from the Azores to Japan, from the shores of Africa to South America, off Canada, on the coast of New England, they hunted the whale. They hunted him in open rowboats, in sailing craft, and in fast steamers; with hand harpoons, nets, blubber knives, spades, prickers, pickaxes, coshes, and grapnel. They hunted him with poisoned arrows, leaving him to suffer for days, finally to die and be washed ashore. Or, making short shrift, they slaughtered him with explosives. A bomb lance patented in the 1850s was fired from a gun: filled with gunpowder, it was designed to be exploded by a fuse in the stem only after it was inside the whale.

Whaling was cruel and ritualized, almost like bullfighting. Here is a description written in 1829 by a Japanese writer, Yosei Oyamada, as quoted by L. H. Matthews in *The Whale:*

> All of the men row near it, striving with each other to be the first, and the catchers stand at the bow, with their harpoons in readiness. After the first and second large harpoons are fast, flags are hoisted at the stern, and the others harpoon the whale in succession. As the large harpoon is made of soft iron, and has its head barbed, it does not break if it is bent, and it does not draw out if it is pulled. And so the whale cannot get rid of it in any way. The whale is weakened by many harpoons, and is in pain, and its moaning is heard like thunder; the surface of the water all around becomes bloody and the red seas run high. The harpooners' boats are towed by the whale as it runs away, and follow it so that the whale is harpooned by them as often as it breaks surface. Then comes the scene for which many catchers are waiting—the lancing—accomplished by the thrusts from lances into the body of the weakened whale; it is terrible enough to make one break into cold sweat.

Robert Cushman Murphy, another writer quoted in *The Whale,* describes the hunting of a sperm whale in 1912, just north of the equator and about midway between Africa and South America:

> We gave our boat-header his opportunity and the horrible lance found the whale's life. "His chimney's afire!" chuckled the heartless mate when the spout which had formerly been thin and white, reflecting rainbows in the late sunshine, became first pink and then thick with gouts of blood. At last we rested our weary selves on our oars, our hands scraped raw by burning rope. Some one

passed out hardtack, and we watched the sun set in the distant rim of ocean
as the whale thrashed into his flurry and turned fin out.

The bloody romance of the nineteenth century gave way to the cold steely horror
of the twentieth. New whaling implements—introduced mostly by Japan, which
followed Norway as the leading whaling nation, to be surpassed in turn by the Soviet
Union—include electric harpoons, rocket-launching devices, helicopters, and sonar
devices that indicate the whale's swimming direction and distance from the ship. The
most modern whalers are 200-foot-long diesel-powered ships weighing 500 to 900 tons
and attaining speeds up to 18 knots. The cannon, placed over the ship's bow, is
loaded with a harpoon filled with blasting powder that explodes inside the whale.
The stricken whale is winched to the surface alongside the ship and a windpipe is
stuck into him, bloating him so that he floats. Then a flag is implanted in his huge
body and he is cut adrift while the catcher goes after the next victim. After the day's
work a floating factory picks up the raw material for processing. The factory ship
is equipped with laboratory, flensing deck, blubber hacker, rotary blubber boiler,
oil separators, bone saws, bone boilers, rotary meat boiler, meat-meal sack-filling plant
and meat refrigeration plant, besides storerooms, recreation rooms, hospital, and
so forth.

The whaling industry reached its peak in the 1930s and 1940s. In 1964, when the
industry was already in decline, there still were 357 whale catchers operating from
23 floating factories, mostly in the Antarctic, and from 39 shore stations in various
parts of the world. The total whale catch was 63,001, including 318 humpback, 372
blue, 19,182 finback, 13,690 sei, and 29,255 sperm whales. The total produce consisted
of 371,413 tons of whale oil, 339,045 tons of meat meal, and other products for food
and fertilizer.

Clipper ships and whalers revolutionized the way fishermen felt about their
occupations, altering the manner in which family and community took part in what
had now become an industry. But the rapid technological development of the fishing
industry was the cause of its own undoing. Ruthless competitive overfishing, combined
with pollution of vast stretches of water caused by the industrialization of the land
and the misuse of its resources, plus the extension of similar malpractices to ocean
space and resources, began to reduce the most important stocks of fish in all the oceans.

In the century from 1850 to 1950 the world fish catch increased tenfold, at an
average rate of about 25 percent per decade. It doubled again in the decade from 1950
to 1960 and again from 1960 to 1970. Thus the protein supply from the sea would
seem to have increased much faster than the world population. In reality, however,
things went differently. The quantitative increase of the catch was accompanied by

a qualitative change. Eventually it was not so much fish for human consumption that were taken in larger quantities as it was plankton-feeding fish swimming in large, tight schools. Previously it had been impossible to fish these species economically; species disdained by tradition and taste, such as very fatty fish; and species that had become useful only with the development of industrial processes that could transform them into fish meal for pig and poultry feed and for fertilizer.

About 35 percent of the world's total catch now consists of "inferior" fish of this sort. But since it takes about ten tons of fish to produce one ton of livestock, as calculated by Sidney Holt, Director of the International Ocean Institute, clearly the real increase in the amount of animal protein available to the increased world population is far less substantial than the statistics appear to suggest.

People in the rich nations fill their animal protein requirements by eating meat and drinking milk—two-thirds of the world's meat and milk production is consumed by less than one-quarter of the world's population. The other three-quarters depends on fish for the greater part of its animal protein.

Nor are the fish-meal products consumed where the need is greatest, in the developing countries. They go to the rich countries, which also have the greater share of the world's fisheries. Over 75 percent of the world's total catch is fished by fourteen nations, including the Soviet Union, Japan, Spain, Poland, France, Norway, and the United States.

In 1970, development began to slow down. Then decline set in. Two dramatic events have contributed to, and are illustrative of, the recent decline of the world's fisheries. One is the shrinking of the whaling industry in the Antarctic and elsewhere; the other is the collapse of the Peruvian anchoveta industry.

As recently as 1937, the various species of baleen whales and sperm whales yielded as much as 16 percent of the world's total marine catch. By 1970 the cruel depletion of these slowly reproducing creatures had reduced the catch to 2 percent. It is likely that even this low percentage, corresponding to 1 million metric tons of whale a year, will disappear over the next decades. It will disappear either because a continuation of overkill will extinguish the species, or because we stop hunting them. There is increasing recognition of the fact that whales, porpoises, and other marine mammals have an almost human intellective capacity, and are of great ecological and evolutionary interest as well. Perhaps both motives—economic failure and humane aspiration—will converge, as they so often do when "progress" is achieved in history, to make whaling a thing of the past. A beginning, at least, was made by the 1975 decisions of the International Whaling Commission, which imposed selective moratoriums on whaling, effectively reducing the world catch by 10,000—from 37,000 to 27,000.

The fall of the Peruvian anchoveta fishery was as precipitous and unpredictable as had been its rise. From a mere 30,000 tons in 1953, the catch of this small fish in the Humboldt Current off the coast of Peru rose to 12 *million* tons in 1970. Small coastal villages grew into huge, bustling fishing ports surrounded by shanty towns that tried to absorb wave upon wave of migrants coming from the country in search of work. The number of fishermen increased from 2,500 in 1957 to as many as 20,000 in 1966. Most of them did not have water, sewers, or electricity in their slums. What they had instead was the stench of fish meal from the factories that produced more than 2.3 million tons of it a year. But they could not afford the meal itself.

The small wooden boats—purse seiners—of the traditional Peruvian fishermen lost in the uneven match with the large, light, fast steel boats of the big companies. In a desperate attempt to meet this competition, many of the small shipowners wrecked their boats by overloading, and many lost their lives. To salvage something by collecting the insurance, many others wrecked their boats intentionally.

Scientists from the United Nations Food and Agriculture Organization (FAO) warned that the huge catches were beyond the sustainable yield, but their warnings were not heeded. Then, in 1972, the fish began to disappear. From 10 million tons the year before, the catch dropped to 4.4 million tons, bringing the world fish catch down by 10 percent. In 1973 the anchoveta catch went down again, to 2 million tons. Environmental factors, such as changes in the Humboldt Current and in the temperature of the water, together with excessive fishing, apparently caused the collapse. Birds which had thrived on the anchoveta—millions of them—disappeared with the fish. Their droppings, called guano, which had accumulated on some of the offshore islands to a thickness of 150 feet and constituted the world's best and most abundant natural fertilizer, disappeared also. And 20,000 fishermen and 8,000 fish-meal workers were jobless.

In 1974, aided by sensible government measures, the anchoveta catch revived somewhat, climbing to just under 4 million tons. It continues to hold its own.

These are dramatic instances, but they are not isolated. As in many areas of human endeavor, there has been tremendous technological change in the fishing industry. And, as in many other areas, technological change unaccompanied by corresponding change in the social and institutional structure has not solved problems but has aggravated them, making the rich richer and the poor poorer. Without protection, the ocean's living resources, the common property of mankind, cannot survive the technological shift from artisanal to industrial fisheries. Nor will the fish heed man-made laws, laws that expand national jurisdiction over the oceans from the traditional three miles to six, or twelve, or fifty, or even two hundred. Fish migrate across the seas regardless of political boundaries.

Bigger and bigger ships, with more sophisticated and costlier gear, are depleting a supply that once seemed boundless. Irrationality is now a structural part of the system. Instead of reducing the catch or restraining the technologies, irrationality triumphs with ever more gadgetry. Hydrophones make high-fidelity recordings of fish noises, which are subsequently analyzed by communications specialists who isolate the bait-eating noises, which are then played back by powerful sound projectors to lure fish over wide distances. Remote telemetering systems listen to fish calls and report suitable concentrations of fish. Completely computerized and automated fishing fleets go into action only when triggered by computers programmed to respond to the telemetric buoys. This is not science fiction. In the near future, as Brenda Horsfield and Peter Bennet Stone report in *The Great Ocean Business,* the computer of a Japanese company will "give instructions to each automatically operated fishing boat, that is, a robot ship. Complete with F.M. automatic fish detector, automatic steering devices, automatic net control systems, and various electronic equipment, the boat rushes to the instructed location. There it determines the specific location, depth, and quantity of the fish involved and automatically selects the most suitable catching method, which would be a fully automated dragging suction pump or an electric fishing net which utilizes the field effect of electric current."

Yet, with all this, the world's fisheries catch less than they could have caught with fewer ships, in fewer hours, and at less expense.

In the salmon fisheries of the United States and Canada, the same annual catch —and total revenue—could be achieved with about $50 million less capital than is currently invested each year. By 1965, the same amount of codfish could have been taken in the North Atlantic with 10 to 20 percent less fishing effort than was employed, at a saving estimated at $50–$100 million a year. The unfortunate Peruvian anchoveta fleet acquired such overcapacity that it could have harvested the equivalent of the annual United States catch of yellow-fin tuna in a single day, of salmon in 2½ days. The Peruvian shore factories overextended their facilities to such a degree that they could have processed the total fish catch of the whole world.

Another striking example of irrationality is offered by the Mexican shrimp fishery in the Gulf of California. In 1950 there were 100 boats, each catching about 100 tons of shrimp a year, the total yield thus being about 10,000 tons. With the development of this fishery, the total yield rose to 20,000 tons a year by 1960; there it stayed until 1970, when it began to decline. The number of fishing boats, however, increased to 400 in 1960—so that each boat now took only 50 tons per year. In 1970, the number of boats had increased to 800, reducing the catch per boat to 25 tons a year. By the middle of the decade, the number of boats was still growing, and the catch was well below 20 tons per boat per year.

This is obvious economic insanity, as insane as the arms race. And it is easy to predict that controls and limitations agreements for fishing will be just as ineffective as arms control attempts have been so far. Therefore the whole basic system of harvesting the fruits of the sea must be changed.

The system *is* in fact changing. The technologies are at hand, and what could be achieved for the benefit of all peoples is so radical that one might call it the Blue Revolution. The scientist-fishermen of this Blue Revolution of sea farming are already at work. If the revolution succeeds, the fishermen of tomorrow will no longer be hunters in the wild. They will be trained in the marine sciences and under their care the ocean will become a nursery for living resources.

The potential of sea farming is stupendous. If the technological changes arising from it are properly harnessed, the Blue Revolution can generate food for the rapidly growing world population.

Sea farming has historical precedents much deeper in the past than is generally assumed. Oysters were cultivated in the Orient long before the Christian era. In the Western world the first oyster farmers were probably the Romans. The Chinese developed a complex ecological fish-culture system over 1,000 years ago. They introduced a number of different species into an aquatic system, each occupying a different habitat and consuming different food. Thus, six varieties of Chinese carp could coexist in one pond: the grass carp, which consumes the large surface vegetation; two mid-water dwellers, one feeding on phytoplankton, the other on zooplankton; and three bottom dwellers, feeding on mollusks, worms, and the feces of the grass carp. This kind of ecological cooperation, with the bottom dwellers absorbing the "pollution" of the grass carp, is far more efficient than any monoculture system.

It is interesting that this experience—or perhaps one should call it this philosophy —goes back so far in Chinese history. It is part of a heritage that makes the Chinese excel today in the design of industrial systems in which one factory uses the waste product of another, and where what they call the "three evils of pollution," namely waste solids, waste fluids, and waste gases, are converted into "three advantages"— resources for new production. According to the Chinese, pollution can only be eliminated when waste is eliminated. Their model is organic and it is ancient.

Primitive forms of aquaculture, both in seawater and in brackish water, have been part of the traditional economy of Southeast Asia for many centuries. Farm ponds are constructed by clearing mangrove swamps and diking them with mud slabs. The ponds are then stocked with fry of various kinds, most commonly milkfish, mullet, and shrimp. Initially they are fed in a "nursery pond," while a community of algae, bacteria, worms, and other plankton is raised in the adjoining ponds. When the fry attain fingerling size, they are transferred to these production ponds, where they mature within a few months or, at most, a year. The average yield of such ponds is about 500 pounds per acre, which compares well with protein production on land.

John Ryther of the Woods Hole Oceanographic Institution made some encouraging extrapolations—assuming that such a simple and inexpensive system could be applied on a global scale. "On the basis of my own calculations, I estimate that there are about one billion acres of coastal wetlands in the world. As a standard of comparison, some seven to eight billion acres of earth are now used for food production, with half of that area devoted to agriculture and half to grazing. If only one-tenth of the available wetlands, or 100 million acres, were set aside for aquacultural development, the potential yield, using improved methods of production, would be 100 million tons a year—the equivalent of the potential yield from the world's commercial fisheries. This rate of productivity is particularly impressive given the fact that it can be achieved through a relatively simple process that requires no extraneous feeding and very little labor or capital investment."

Ryther's estimate is realistic—one might even say conservative. According to the calculations of Horsfield and Stone, if it were possible to turn just the 1,000 square miles of Long Island Sound over to mussel culture, this area could produce a quantity of protein equal to three times the total world fish catch. It should be added, however, that the selection of wetlands for fish farming must be preceded by the most careful ecological investigations, for we know that wetlands, especially mangrove swamps, are the nursing grounds of most marine animals that later migrate out to sea, and must not suffer interference.

Some successes have already been recorded. When the oyster was overfished in the nineteenth century and became scarce, the French began farming it. Now the Bay of Arcachon, near Bordeaux, produces about 500 million oysters every year for the European market. Japan, which derives as much as 13 percent of its total ocean produce from mariculture, has raised the productivity of oysters from 600 pounds per acre under natural conditions up to 32 tons per acre under culture—a hundredfold increase.

Green turtles are farmed by Mariculture Ltd. in Turtleland, on the Cayman Islands in the Caribbean. The ten-acre farm consists of a system of about 160 concrete pens and tanks, a huge artificial breeding pond containing a million gallons of seawater, a nesting beach, processing facilities, laboratories, and offices. More than 2.6 million gallons of seawater are circulated through the pens and tanks every hour by a network of pipes. According to the company's statements, about 160,000 turtle eggs, partly "home-grown," partly collected from nesting beaches, are hatched every year at Turtleland. Whereas in nature probably fewer than two hatchlings in a thousand survive, Mariculture Ltd. is obtaining a survival rate of about 95 percent.

The British Oxygen Company has designed a special tank in which up to seven rainbow trout can be raised per cubic foot of water, which is then recycled at a very low loss of about 10 percent. In nature only one trout can be raised per cubic foot of water, and ten gallons a day of fresh water are required to produce one hundred tons of trout a year.

Kelp and seaweed are being cultivated for human consumption—albeit as condiments—and for a growing number of industrial and pharmaceutical uses. Carp, mullet, and milkfish, plaice and whitefish, trout and salmon, shrimp, squid, and abalone are already being farmed. And about 3,000 tons of farmed sea urchin roe were consumed in Japan last year.

A most beautiful and complex system, inspired by the ancient Chinese polyculture, has been devised by John Ryther at Woods Hole. Here sewage from a nearby secondary treatment plant is used to grow plankton algae, which in turn provide food for shellfish, principally oysters. The algae remove nutrients such as ammonia, nitrate, and phosphate from the sewage effluent, and the oysters remove the algae from the water. The oysters return some of the nutrients to the water in the form of excreted wastes. These are consumed by seaweeds, especially sea lettuce, which are added to the system for this purpose. The seaweeds are then fed to abalone. The oysters' solid wastes, which drop to the bottom of the tank, are eaten by sand worms, which are then circulated to a neighboring tank to serve as food for flounder. The products of this continuous culture system are a primary crop of oysters, side crops of seaweeds, worms, flounder, and abalone, and, ultimately, clean water which is returned to the sea. The sea farm in fact becomes a tertiary (or biological) sewage treatment plant. If implemented on a large scale, Ryther concludes, "such a system would be capable of producing an annual crop of one million pounds of shellfish meat from a one-acre production facility and a fifty-acre algae farm using effluents from a community of 11,000 people. The potential yield of world-wide aquaculture, based on the simplest improvements, is already an impressive 100 million tons of food. By adopting advanced culture techniques such as that developed at Woods Hole, the yield could well be multiplied tenfold within the next three decades."

Of course, there are as many problems as there are fish farms. On the biological side there are parasites, blights, epidemics. On the economic side there is the relatively high cost per unit in many cases. But the basic principle is the same throughout: to take advantage of the lavishness of nature and raise the survival rate of eggs and larvae—tenfold, fiftyfold, a hundredfold. Once the fish have been raised to the critical fingerling size, they can be released to a natural environment, provided it contains sufficient nutrients, whether natural or introduced by man.

We are only at the beginning of this development, only on the margins of the ocean. But the Blue Revolution need not stop at the water's edge. It can be carried into the deep seas. Whole ocean areas—the Mediterranean and the Caribbean—could be turned into fish ponds. The productivity of the water could be multiplied. A Soviet scientist, Boris Bykhovsky of the Academy of Sciences, has been working on schemes to improve the productivity of wide ocean areas by controlling the transfer of substances and energy. He thinks that new varieties of single-celled algae could be developed which would utilize the sun's energy more effectively. Thus even migratory fish could be "cultured"—raised to viable size in the desired quantity in hatcheries—and then released into wide-open fertilized spaces where there is as yet less competition with other users of ocean space than in the congested coastal areas.

The present pattern of distant-water fishing is likely to change under the impact of these developments—combined with a number of other factors, such as the extension of national sovereignty over wide coastal zones and the increase in fuel costs. Fisheries may develop in three new directions. First, fish farming of the kind just described. Second, the catch of very small organisms, such as plankton and krill, may be multiplied, and these may be processed into food for human consumption. The Soviet Union is pioneering in developing such processes. The abundance of Antarctic krill alone is such that its systematic harvesting and processing could more than double present food production from all the oceans. Third, new mid-water trawling technologies, combined with new food-processing methods, make available for human consumption a wealth of mid-water creatures. One of these alone—squid—also exceeds in quantity the total present world fish catch. In these technologies the Soviet Union leads again at this time.

In the light of such possibilities, does it not seem atavistic to continue to exterminate the whale, or even the anchoveta? Does it not seem futile to haggle over boundaries? The ocean is a polyculture whose manifold components must be understood and managed. Its uses, now in conflict, can instead be integrated. We have the tools, such as systems theories and computer sciences, to do it. We can turn wide ocean stretches into fish farms and manage them cooperatively, not competitively—some on a local, some on a regional, some on a global basis. Obviously there are great problems in projects of such scale, which require so much capital, technical skill, space, and labor. It is a huge task, but it is a *positive* task. And we can do it *if* we do it together.

They that go down to the sea in ships
. . . these see the works of the Lord,
and his wonders in the deep.

Psalm 107: 23–24

CHAPTER FIVE

THE SAILORS

THE FIRST ROAD WAS A RIVER, and it was very well engineered. How it curved its way around obstacles or dug its way through rocks!

Rivers were the arteries of ancient kingdoms, the natural links between village and village. "I travel," in the ancient Egyptian language, could only be said by "I sail downstream" or "I sail upstream." Sailing downstream, the river is the road and the power that moves us along the road. When we build a moving sidewalk, we merely make the road more like the river, its pristine model.

People cut wood in the hills upstream, wood they needed down by the oceanside, where the fish were. Then they floated the logs down the river. They could have learned this from the animals: beavers make quite a science of cutting timber and shipping it downstream. With their dams they even regulate the level of the water so that the timber can travel smoothly.

Eventually people helped their timber move along by tying a few logs together and traveling on them to steer them clear of impediments. Then they began to load aboard the mushrooms and wild berries they had picked on the hillside. Why carry them if the river will do it for you?

Thus, simply stated, the first ship was born.

Other objects also floated and could be carried by the river: wooden tubs, pottery, baskets, hollow halves of bamboo stems, gourds and other fruit, papyrus and other reeds, bloated animal skins. Many boats bore traces of their origins from such objects for thousands of years—some still do today.

The earliest pictures of boats are Stone Age rock carvings in Norway. Many more have been found in southern Scandinavia, dating from the Northern Bronze Age. They are very simple, like drawings made by children.

Egyptian boats, or rather rafts, of the Predynastic period (between 4500 and 3000 B.C.) were already remarkably sophisticated. Made of papyrus, they had many oars, like cilia, and rudders like the web-toed, back-stretched feet of waterfowl. They had deckhouses or cabins, and anchors, and carried palm trees to give the oarsmen shade. The palms caught the winds in their slender leaves, which could be braided or matted so that the wind would push them.

Thus the first sail was born—at least 6,500 years ago.

Ships reflect—sometimes even anticipate somewhat—a people's stage of civilization. Nautical construction and the art of navigation are significant clues to a culture.

Consider what goes into the building of a ship: the whole arsenal of a people's crafts and sciences, of art and life style, of world view and collective purpose. A ship epitomizes man's attitudes toward other men and toward nature.

The evolution of the hull, the water house, is basically simple, very slow, and amazingly universal.

The Egyptian papyrus boat, originally just a bundle of reeds with their ends tied and turned up, determined the shape of generations of wooden boats in many parts of the world. The Chinese model, derived from a split bamboo stem with its divisions, which became transverse bulkheads, influenced European shipbuilding both in the Mediterranean and in the North. The dugout canoe, invented wherever there were logs, fire to burn them out, and flints or adzes to remove the burned-out insides, developed into a keel by the addition on either side of planks or strakes. These were either laid overlapping, giving rise to generations of the clinker-built boats typical of Northern Europe; or they were abutted edge to edge, forming the smooth external and internal surfaces of carvel-built ships. The ships of the European Mediterranean, of the Persian-Arab civilization, of India, Oceania, and China, were all carvel-built.

The strakes might be sewn together with primitive ropes or nailed with wooden pegs or pins. Frames and ribs might be constructed first and then covered with strakes, in a line of evolution converging with that of the primitive skin boat of the Stone Age Scandinavians and Eskimos, the caulked floating baskets, such as the *quffa* and *hisbiya* of Iraq, or the mud-caulked baskets of Indochina. The latest version of this ancient model is the iron-cement fishing boat of Bangladesh, shaped of chicken wire and caulked with cement—a process applying modern technology to an ancient pattern, and so simple that any fisherman can do it in his backyard.

Or the planks may be fitted first and a frame inserted subsequently, to reinforce the structure, as may become necessary when boats originally built for river navigation are launched into the open sea. This was the way the ancient Nordic people built their boats throughout the Viking period. But whether the Viking ship derived from the log canoe, to which planks were added, or from the skin boat, with the planks substituting for the skins, we do not know.

The keel, furthermore—absent from the Chinese bamboo-originated junks and sampans and from the flat-bottomed boats of Central Africa—had yet another ancestor. Some of the boats of India may have derived from wooden rafts or catamarans (from the Tamil *kattumaram,* "tied logs") rather than from dugouts. On the Malabar and Coromandel coasts, Joseph Needham notes in his monumental work on nautical technology in China *(Science and Civilization in China),* "there is a tendency to build rafts of odd numbers of logs so fixed that the central one is the lowest, thus approximating to a keel."

While there was little basic change in the way boats were built from prehistoric times to about A.D. 1200, forms had mingled a great deal by the time of the First Crusade (1095–99). Some of the similarities between ships from very different cultures may be ascribed to a general mingling of cultures—the world of oceans and marine ports has always been the most international. Others result from "convergence"—that is, identical traits evolving independently from quite different origins. Thus James Hornell *(Water Transport)* points out amazing technical parallels in the hull construction of Scandinavian and Oceanic boats: "Certain boats in that region still in use not only look like Viking longships but have the identical method of lashing the strakes by cleats to transverse ribs." English boats, like the English language, have a dual origin, some elements deriving from Norse, some from Mediterranean influences. American Indian sailing craft underwent influences from China, probably in the wake of very early crossings of the Pacific from Asia to America, and remarkable similarities exist between Formosan and Ecuadorian-Peruvian forms. And Chinese and Arab, Nordic and Mediterranean features merged to produce the swift three-masted ships which, in the fifteenth and sixteenth centuries, circumnavigated the globe.

Naval technology is a world language. Yet there remains in naval architecture, just as in architecture or art in general, much that is particular and deeply rooted in specific world views. Thus Needham points out that, broadly speaking, the European tendency has always been toward setting the greater fullness of a ship toward the bow, while the Chinese tendency is to set it toward the stern. This he roots in a conceptual difference—the choice of a model in nature. While the Europeans believed they should construct their boats with the outline of a fish, the Chinese felt they should follow the outline of a bird swimming. One of the earliest to stress this fundamental difference was the French Admiral F. E. Paris, whom Needham quotes as having written, in 1840:

> For our best hulls we have taken the fishes as models, always larger at the cephalic end, but the Chinese, who also copied nature, imitated the palmipeds, which float with the greatest breadth behind, for somewhat obscure reasons. In this they were very acute, for aquatic birds, like boats, float between the two media of air and water, while fish swim only in the latter. These strange people seem to have done everything in the opposing way to which it is done at the other end of the continent, and they copied Nature still further in seeking to exert the greatest possible propulsion on the stern, instead of applying, as it were, a tractive force to the prow. This led them to the use of those strong paddles (the Yulohs), which imitate in position the web-feet of the palmipeds— a position which must have been very important for swimming, since it deprived such birds of the facility of walking easily on land and even, in the case of the best swimmers, prevented it altogether. These very simple observations (which the Chinese have utilized) will find one day, perhaps, a happy application to the steam-boat, which, set in motion as it is by an internal force not coming like the wind from without, finds itself in exactly the same situation as the swimming bird, and might gain from a closer approximation to the latter's form.

Paris's remarkable prediction, as Needham points out, was realized within a couple of decades with the invention of the screw propeller. The Industrial Revolution then standardized and universalized the shape of ships as it universalized and standardized so many other things.

If the hull—birdlike or fishlike, open or decked, continuous or compartmentalized, with deckhouses and towers or without—was the body, the energy that moved it was the spirit. Body and spirit, hull and propellant of course interacted, and so, too, did the history of ship construction with the evolution of energy technologies and resources. Sometimes it was ship construction that led the rest of civilization in the conquest of new levels of energy; sometimes ship construction followed originally land-oriented developments.

The oldest form of energy, obviously, is muscle power, whether animal or human. Animal power was practical on land, where it carried its own weight, but not on water, where it preempted too much space and loading capacity. Primitive navigation thus depended on human power, the first and simplest extension of which was, universally, the paddle. Boats everywhere were paddled before they were rowed. The oar, technologically a far more efficient device, producing more power with less muscle force, had to await the invention of the fulcrum and its transfer from land to water. There is evidence from rock carvings that boats were being rowed in Scandinavia in about 300 B.C. In Egypt the oar was introduced far sooner, certainly by 3000 B.C.

In the East, the use of wind power through the sail preceded oar power. The oldest Egyptian square sail goes back to 6000 B.C.; it spread from Egypt to the Mediterranean, India, China, even to the Vikings. Thus, wind power was first harnessed by navigation. It is curious how very long it took to harness it for land use. The earliest known references to windmills are to a Persian millwright in A.D. 644 and to windmills in Seistan, Persia, in 915.

Waterpower was also first used by navigators. The discovery of waterpower almost forced itself on the boatman floating his primitive craft down the river road, but its use on land had to await the invention of the wheel, a development first applied to muscle-powered locomotion (Mesopotamian carts, about 4000 B.C.) and then to the waterwheel (Roman and Greek, first century B.C.).

Waterpower, the first nonmuscle power to be applied, is an amazingly complex power. Needham distinguished two kinds of waterwheels: ex-aqueous and ad-aqueous. Ex-aqueous wheels are driven by the force of falling (or flowing) water; ad-aqueous wheels are driven by another force which is applied to moving the water. The watermill is based on an ex-aqueous wheel, the ship's paddle wheel is ad-aqueous. In the same way the windmill is ex-aerial, the plane's propeller ad-aerial. Only the ex-aqueous wheel uses waterpower.

Water has another form of power, however, and that is its lifting power, based on its specific weight and density. This is experienced by the swimmer when he lifts his feet from the ground and floats.

A packhorse, according to A. W. Skempton ("Canals and River Navigation Before 1750") can carry a quarter of a ton on its back. When harnessed to a wagon, it can transport considerably more: on a soft road it can pull a little over half a ton; on a macadamized road it can pull two tons, and if the wagon moves on iron rails, it can pull as much as eight tons. But if hitched to a barge, a packhorse can pull thirty tons on a river, fifty on a canal—that is, 400 times what it can carry on its back! Such is the power of water.

The use of water's carrying power, the use of muscle power applied to paddle and oar, and the use of wind power applied to the sail characterize various phases in the evolution of navigation. And here, again, there are important cultural differences. For instance, Europe clung more tenaciously to muscle power, the development and exploitation of which, in the huge slave-rowed galleys of Roman and medieval times, engendered some of the most inhumane scenes the oceans have ever witnessed. The use of galleys lingered on in the Baltic until the nineteenth century. Their graceful lines and elaborate decorations concealed incredible misery within, for the galleys were rowed by convicts, slaves, and prisoners of war who were condemned to a life sentence at this labor. They sat chained to their posts, usually five to each oar, covered with vermin. They were whipped to spur them on and tossed overboard when they died.

The Chinese never developed the galley at all. Chinese sailors rowed, but the rowing was auxiliary and they did not use slaves for it. They relied on wind power rather than muscle power, and this they could well afford, for until the fifteenth century A.D. their naval technology was far superior to any other in the world.

The main point in harnessing wind power was, of course, to devise a sail or combination of sails that could be used in navigating against the wind, and although right whales were born knowing how to do it, for Homo sapiens it was not so easy. The Chinese solved the problem at an early date: the threemaster, the sprit sail, developed in Europe only in the fifteenth century, was already well established in China in the seventh. The Cantonese had four-masted lug-sail boats even in the third century.

Sails differ basically in their shape, in their attachment to the mast, and in the way they are stretched—some being stretched by one "yard" on top or diagonally, others by two: one at the top, one at the bottom, or one on either side. Basically, one may distinguish fourteen kinds. Sails may be made of woven palm leaves, of animal hides, of homespun wool, of flax or linen, or of synthetic materials such as nylon. They may be stiffened by battens of bamboo, like the unique mat-and-batten sails of Chinese junks.

There may be one or more masts, of different sizes, in one line or staggered. They may stand on one leg or two, or they may be supported by some sort of cage.

The simplest and oldest sail is the square sail, which originated in Egypt about 6000 B.C. and spread all over the world. It can only sail before the wind. To go against the wind, a type of sail is required that can be turned to receive the wind on either side—a fore-and-aft sail.

One of the oldest fore-and-aft sails was the triangular lateen sail (lateen is a misspelling of the word Latin), introduced into the Mediterranean by the Arabs in the ninth century. It dominated the Mediterranean until the thirteenth century. Early in the fourteenth century, a development of great importance took place: the square sail, which was still used by the Vikings, was combined with the Arab lateen sail in a two-master. The Portuguese three-masted caravel of the fifteenth century represents a further phase of this development, under the added influence of the multimasted Chinese boat. Columbus's *Santa Maria* was such a ship. Typically, it carried five sails on its three masts: a bowsprit, two sails, a topsail, and a lateen.

Then, in the sixteenth century and mostly under Dutch influence, mast was added to mast and sail to sail, producing six-masters with nearly fifty sails, combining all shapes and fashions. So complex was the rigging that officers taking their turn at the helm during an ocean voyage were under instruction not to order any changes in the rigging for at least the first half hour of their turn, for it would take even the most expert that long to understand the rationale of the arrangement he took over.

The queen of the sailing ships, combining grace, safety, and speed, was the British and American clipper ship of the nineteenth century. The clippers brought tea from the Indies and immigrants from Europe to Australia and America. They could cross the Atlantic in fourteen days and round Cape Horn, from New York to San Francisco, in less than 100 days.

But by then the Industrial Revolution was under way, and for the first time a land-developed energy source, steam, began to conquer the oceans. (The Chinese, curiously enough, had a steam-turbine-driven paddle-wheel boat as far back as 1671. It was a toy built by a Jesuit, Philippe-Marie Grimaldi [Min Ming-Wo], for the young K'ang Hsi emperor, but it was never duplicated in full scale.)

The new technology did not triumph without resistance. The first two steamboats developed in the West were both sunk by jealous sailors. The destruction in 1707 of Denis Papin's prototype of a ''cooking-pot''-powered paddle-wheel boat may have delayed the take-over of the steamboat by a century. After James Watt's invention of the steam engine, two Frenchmen, Count Auxiron and Count Follenay, tried again, about 1770, but their boat was sunk in the Seine before its maiden voyage.

There were other experiments during the remaining years of the century, in France and especially in America, where the long distances, lack of roads, and abundance of rivers created particular needs and opportunities. The first full-blown and commercial success was Robert Fulton's *Clermont,* which initiated a regular passenger service on the Hudson River between New York and Albany in 1807.

By the middle of the nineteenth century steamboats had conquered all the rivers and lakes of Europe and crossed the English Channel. In 1819 a full-rigged sailing ship, the American *Savannah,* equipped with auxiliary steam-powered paddle wheels, made the trip from Savannah, Georgia, to Liverpool, England, in twenty-nine days. The steam engine was used for ninety hours during the voyage.

Steam power went together with wind power, as wind power had gone together with muscle power before, and thick, black, smoking chimneys took their place between the slender masts with their white sails.

Regular steamship passenger service across the Atlantic was inaugurated in 1838 by the Great Western Company, followed by the Cunard, Hamburg-American, Norddeutsche Lloyd, Holland-America, Red Star, and French lines. Wooden hulls gave way to iron and steel hulls, paddle wheels to the screw propellers. The tremendous *Great Eastern,* designed by Isambard Kingdom Brunel and launched in 1858, combined everything: five funnels, six full-rigged masts, two 58-foot paddle wheels, and a 24-foot screw propeller. She was 680 feet long, displaced 22,500 tons, made 25 knots per hour, and carried 4,000 passengers.

In line with the accelerating tempo of the Industrial Revolution, steam engines passed through various stages of refinement: from boilers with external firebox to boilers with internal firebox; from fire-tube boilers to water-tube boilers. Greater efficiency was achieved by the introduction of a three-cylinder engine with high-, medium-, and low-pressure cylinders. The steam engine gave way to the steam turbine, transforming energy directly into rotation. The steam yacht *Turbinia,* constructed by Sir Charles Parsons, was the first embodiment of this new process. After some trial and error, building and rebuilding, *Turbinia* reached the incredible speed of 34 knots and put to shame the most powerful ships of the British Navy, whose Admiralty quickly adopted the new invention.

Then came, in short order, the diesel engine and the diesel turbine. Now the Atomic Age seems to have opened a new epoch in the history of shipping. Simple ships they are, really, the nuclear-powered submarines and cargo vessels, much simpler, in a way, than the glorious clippers. They are nothing but steam turbines, but the fission of one unit of nuclear fuel produces two million times as much heat as the burning of one unit of fuel oil, and one can cross the Atlantic with a tablespoon of fuel!

But hull construction and energy generation are not all that goes into shipbuilding and navigation. Navigators must take their bearings—longitude and latitude—far out of sight of land, under clear or overcast skies, by day or night. They must steer their vessels and anchor them; they must fathom the depths over which they sail, lest they run aground on sandy shallows or tear their bottoms open on rocks and reefs. They must know currents and tides and landmarks. They must be able to predict weather and winds. How the ancient seafarers, whether Chinese, Phoenician, or Viking, navigated in the wide oceans, without science or instruments, except for a lead line and, perhaps, a wind rose, is mysterious—as mysterious as the migrations of eel, homing salmon, or green turtle.

Needham distinguishes three great periods of navigation. The first was a primitive period, which in Europe lasted until the year 1200 and in East Asia, which then was far ahead of the rest of the world, until 900. The second was the "quantitative" period, which could also be called "protoscientific." It lasted until roughly 1500. Third came the "mathematical," or, more broadly, the "scientific" period.

In their way, the "primitive" sailors knew a great deal. Needham quotes a text from the fourth century A.D. which describes the skills of a famous Indian pilot named Suparaga: "He knew the course of the stars and could always readily orient himself; he also had a deep knowledge of the value of signs, whether regular, accidental, or abnormal, of good and bad weather. He distinguished the regions of the ocean by the fish, by the color of the water, by the nature of the bottom, by the birds, the mountains [landmarks] and other indications." And Bjørn Landström, in *Ships of the Pharaohs,* quotes another testimony to early nautical meteorology in the "Tale of the Shipwrecked Sailor," which took place during the reign of Amenemhat I, of the Twelfth Dynasty: "I ventured out on the big green in a ship of 120 cubits [60 meters] long and 40 cubits broad. One hundred and twenty of Egypt's best sailors were aboard. They looked to the sky, they looked to the land, and their heart was braver than the lion's. They foresaw a storm before it had come, and a tempest before it struck."

"Primitive" sailors took soundings, gathered sea-bottom samples, marked the prevailing winds and currents, recorded depths, anchorages, landmarks, and tides. "Primitive" ship construction produced the fleet of ships Queen Hatshepsut of Egypt dispatched to Punt about 1500 B.C. An impressive assembly of ships it must have been. Each one was about seventy feet long, with an eighteen-foot beam. More than ordinary cargo vessels, they were trading galleys built for fast voyages in dangerous waters. They were built with keels and with stemposts and sternposts. The structure was reinforced by ropes (hugging trusses). The mast, on which the square sail was hoisted, was probably two-legged. On a relief in the queen's burial temple in Deir-el-Bahari we see fifteen sailors standing on their long oars on each side of the ship, and others standing on the bridge or climbing around on the sail's boom. There were two stern rudders, like duck's feet, to steer the vessel, and a beautifully decorated castle, with cabins for the ship's officers in the rear. Säve Söderberg, in *The Navy,* describes the open deck loaded with merchandise, which included "many precious woods from the land of the god, together with quantities of fragrant gum and fresh incense, trees, ebony, ivory, pure gold from Amu, fragrant wood, kesit wood, ahem incense, holy gum, eye paint, dog-headed apes, long-tailed monkeys, dogs, and also leopard skins and natives and their children."

"Primitive" seamanship produced the Phoenician galley, which may have circumnavigated the continent of Africa over 2,500 years ago. The Phoenicians built their ships of cedar of Lebanon, and traded in silver, lead, and iron, traveling to

Malta, Sicily, Sardinia, and Spain. They passed the Straits of Gibraltar and may even have reached Britain, establishing trading stations all along the way. The construction of the Homeric ship, which was contemporary with Phoenician trading ships of about the seventh century B.C., is described in Robert Fitzgerald's translation of
The Odyssey, Book V (Doubleday, 1961):

> When Dawn spread out her finger tips of rose
> Odysseus pulled his tunic and his cloak on,
> while the sea nymph dressed in a silvery gown
> of subtle tissue, drew about her waist
> a golden belt, and veiled her head, and then
> took thought for the great-hearted hero's voyage.
> A brazen axehead first she had to give him,
> two-bladed, and agreeable to the palm
> with a smooth-fitting haft of olive wood;
> next a well-polished adze; and then she led him
> to the island's tip where bigger timber grew—
> besides the alder and poplar, tall pine trees,
> long dead and seasoned, that would float him high.
> Showing him in that place her stand of timber
> the loveliest of nymphs took her away home.
> Now the man fell to chopping; when he paused
> twenty tall trees were down. He lopped the branches,
> split the trunks, and trimmed his puncheons true.
> Meanwhile Kalypso brought him an auger tool
> with which he drilled through all his planks, then drove
> stout pins to bolt them, fitted side by side.
> A master shipwright, building a cargo vessel,
> lays down a broad and shallow hull; just so
> Odysseus shaped the bottom of his craft.
> He made his decking fast to close-set ribs
> before he closed the side with longer planking,
> then cut a mast pole, and a proper yard,
> and shaped a steering oar to hold her steady.
> He drove long strands of willow in all the seams
> to keep out waves, and ballasted with logs.
> As for a sail, the lovely nymph Kalypso
> brought him a cloth so that he could make that, too.
> Then he ran up his rigging—halyards, braces—
> and hauled the boat on rollers to the water.

"Primitive" seamanship produced the Roman merchantman which, for five centuries, dominated the Mediterranean from Britain to the Black Sea, sailed around India, traded with Malaya, Sumatra, and Java, and reached the borders of China. The Roman grain ships were enormous, the "tankers" of antiquity: 180 feet long and with a capacity of 1,300 tons of cargo. "What a tremendous vessel it was," Lucian wrote of one of the sailing ships, carrying grain from the Nile valley and gold from the Nubian mountains to Rome, "the crew was like an army!"

And then of course there were the unsurpassable Viking longships. Their clinker build, with overlapping planks, gave them a unique kind of elasticity on rough seas. Captain Magnus Anderson, who in 1893 sailed a replica of a Viking longship from Norway to Newfoundland, reported that the gunwale would twist half a foot out of line and the bottom would rise nearly an inch, like the chest of a breathing whale. This elasticity gave the ship her strength. The largest longships were over 100 feet long, with shallow draft, high prow and sternposts, and one mast with a broad, strongly sewn and beautifully colored and decorated square sail. They carried sixty or even eighty oars and were steered by a deep oar on the right side of the stern ("steer board," which became "starboard"). The Vikings came down from the fjords and settled Iceland and Normandy, visited London, Hamburg, Paris, Lisbon, and Pisa, and circumnavigated the North Cape and discovered Spitzbergen. They sailed the Black Sea and the Caspian. They passed the Dardanelles and arrived at Byzantium. They discovered America around 1000, sailing from Iceland to Greenland and then down the eastern seacoast as far as Cape Cod or even further.

The second, or "quantitative," period of navigation dates from the thirteenth century A.D. "Measurement was the keynote of this period," Needham says, "when pilots would no longer sail by guesswork and the help of the gods."

The chief instruments of measurement and "dead reckoning" ("dead" being a mutilation of "deduced") that came into use during this period were compass for direction finding; maps; hourglass to measure time and speed; and astrolabe, sextant, or crossbar to measure the altitude of the stars. Trigonometry was used to calculate wind- or current-induced deviations from the established course.

The magnetic compass was used widely on Chinese ships sometime about 1000, quite possibly as early as 850, with its first beginnings reaching back as early as the first century B.C. The earliest description of a floating compass dates from the middle of the eleventh century. It was a thick leaf of magnetized iron with upturned edges, cut into the shape of a fish. How soon use of the compass spread to the Indian Ocean is not known, but we do know that the Arabs had it next, and that it was they who transmitted it to the Mediterranean, where it was first used in the twelfth century.

Curiously enough, in both China and Europe the introduction of the compass roughly coincided with the invention of the sternpost rudder, which was a significant improvement in the technology of steering. It was as though there were a feedback between the capability for orientation and the art of steering.

But orientation, even with the help of a magnetic compass, was of little avail if the pilot could not establish where he was. The use of astrolabe and cross-staff to measure the altitude of stars is documented in China as early as the eleventh century, in Europe by the fifteenth. Archaeological finds in the Mediterranean indicate that there may have been astronomical computers in antiquity.

The first sea chart in Europe is the Carta Pisana, dating about A.D. 1275. The Chinese made maps about 800. The principle of the rectangular grid on which to plot the map goes back as far as the third century.

Time was measured in Europe with hourglasses from 1310 on. The hourglass came from Venice, where glassblowing was first mastered, but may have originated in the East, together with the compass and the sternpost rudder. The Chinese probably relied on a "combustion clock" at sea; that is, they kept incense sticks burning in the ship's shrine and measured the time by how many sticks had burned.

The kinds of fleets produced by the "quantitative" period were, again, quite astonishing. We shall single out two examples, the Chinese fleet from the twelfth to the fifteenth centuries, and the achievements of Prince Henry's Portuguese school of navigation at Sagres in the fifteenth century. This school marks the transition from the second to the third, or "mathematical," period of navigation.

China's first permanent navy was created after the removal of the capital to Hangchow, in southern China, in the twelfth century. From then on, China regarded the sea as her Great Wall and substituted warships for watchtowers. Between 1132 and the end of the century, a great number of treadmill-operated paddle-wheel craft were constructed. They were equipped with as many as eleven pairs of paddle wheels, and were armored with iron plates.

In 1170 a traveler on the Yangtze River observed naval maneuvers in which 700 warships participated, each about 100 feet long, with castles and towers. They sailed rapidly, even against the current. Another observer, Chou Chün, described the sea-going ships of this period in glowing terms: "The ships which sail the southern sea and south of it are like houses. When their sails are spread they are like great clouds in the sky. Their rudders are several tens of feet long. A single ship carries several hundred men, and has in the stores a year's supply of grain. Pigs are fed and wine fermented on board. There is no account of dead or living, no going back to the mainland when once the people have set forth upon the cerulean sea."

Marco Polo's admiration for the Chinese Navy knew no limits. When he left

China in 1292, he wrote: "The Great Khan caused to be armed and set forth 14 great ships, and every one of them had four masts. . . . In every ship he put 600 men, and provision for two years. . . . It was almost certainly a much braver fleet than any European country, including the England of Edward I and the France of St. Louis, could have launched for the occasion." It was this fleet that was to dominate the East China Sea, the South China Sea, and the Indian Ocean for three centuries and, under Admiral Cheng Ho, rounded the Cape of Good Hope.

At its peak, under the Yung-lo Emperor, about 1420, the Ming navy consisted of some 3,800 ships in all, including 1,350 patrol vessels and as many combat ships attached to guard stations or island bases, a main fleet of 400 large warships stationed near Nanking, and 400 grain ships or galleons, manned in some cases by over 1,000 sailors each. There were 3,000 merchantmen ready as auxiliaries, and uncounted small craft such as dispatch and police launches.

After 1433, this magnificent Chinese fleet declined—even faster than it had arisen. It was not defeated in battle, but gave way to a new Chinese policy, that of the Neo-Confucians, whose ambitions were land- rather than sea-oriented. By 1500, it was a capital offense to build a seagoing junk with more than two masts.

As the Chinese went down, the Portuguese came up.

Henry the Navigator, third son of King John I of Portugal and his British queen, Philippa of Lancaster, founded at Sagres what we would call today an international and interdisciplinary ocean institute. His goals were to find a southern passage to India; to outflank the Muslim empire; to enrich trade; and to carry the Christian faith around the world.

His means were the scientific improvement of navigation until it proved adequate to the task.

He gathered mathematicians, astronomers, chartmakers, captains, and chroniclers —Portuguese and Spaniards, Venetians and Genoese, Arabs and Jews. They worked on chart design; they improved compasses, astrolabes, and quadrants; they compiled astronomical tables. And it was they who, in a series of experiments, joined the ancient square sail to the Arab lateen and constructed the first caravel.

The school of Sagres produced not only theoreticians and naval architects, but explorers and conquerors as well. Columbus was a student there, and Sagres gave the world a whole generation of great seafarers, from Gil Eanes, who rounded Cape Bojador in 1433, to Vasco da Gama, who landed in Calicut on the west coast of India in1498. Suddenly, Portugal had snatched maritime supremacy of the Indian Ocean out of Arab hands.

From 1500 onward, as Needham so aptly puts it, "new aids for the sea pilot came tumbling out of the cornucopia of the 'new, or experimental, philosophy' in a wealth almost as bewildering as that of the aerofoil or transistor age."

There was no science from which nautical technology and navigation did not profit during the next half millennium; there was no voyage of discovery that did not make new, stimulating demands on science.

The sixteenth century contributed significant advances in astronomy—including tables and instruments, and the refinement of the quadrant into the sextant. The seventeenth century did not introduce basic changes, but the eighteenth brought decisive improvements in meteorological forecasting with the invention of the marine barometer, and solved the problem of accurately determining longitudes with the development of marine chronometers. It also brought a significant expansion in ocean trade, revolutionizing life styles in Europe. The average size of merchantmen increased fivefold, from 150 or 200 tons in the seventeenth century to 1,000 tons. But their scientific equipment was to remain basically unchanged until World War I. The nineteenth century added thermodynamics and metallurgy, the twentieth saw the birth of electronics. Radio, radar, satellite communication, and computer science have once more revolutionized the art of navigation—by making it, in fact, unnecessary.

Are we, qualitatively, still living in Needham's third, or "mathematical," era of navigation, or is it more meaningful to close this period with World War II? This third period would then include the *Santa Maria* and the *Queen Mary,* the Spanish Armada and the Allied fleet—5,000 ships and landing craft that carried the Allies to Normandy in 1944, the largest assembly of ships the world has ever seen. The third period solidified and amplified the horizontal conquests of Vasco da Gama, Columbus, Vespucci, Magellan, and Cook.

The fourth period, the "nuclear-electronic" period, has more dimensions, because it is the outer-space age and the inner-space age. Its problems—for example, pressure and breathing—are very similar, and so are the technologies applied to their solution. The companies managing the two technologies are often the same, and so are the pioneers and heroes, from balloonist and deep-sea diver Auguste Piccard to astronaut/aquanaut Scott Carpenter, the vertical Vasco da Gamas of our age.

The submarine did not arrive all at once. It had been in the making for several hundred years.

We need not go back to the ancient pearl and sponge divers, or the various diving suits and bells devised for them through the millenniums. Even the self-propelling submersible vehicle has a history almost as long as the "mathematical" period of navigation. The first design for a submersible appears in the sixteenth century, but its inventor, Leonardo da Vinci, kept it secret because he thought it was too destructive.

Robert Fulton, who was so successful in developing the steamboat, also built submarines. His *Nautilus I* and *Nautilus II* were equipped with mast and sail for surface navigation and a hand-operated paddle wheel for underwater navigation. In June 1801 *Nautilus I* was demonstrated on the Seine. It navigated a considerable distance underwater, for a period of 25 minutes.

Quite remarkable was the success of a submarine built in France in 1846, which James Sweeney describes as the first purely scientific submarine *(A Pictorial History of Oceanographic Submersibles)*. Sweeney quotes the following description from the *Illustrated London News* of 1854: "The inventor, Dr. Payerne, has not only discovered means to descend to the bottom of the sea and to work there at his ease with a body of operatives, and to remain there as long as he pleases, replacing by chemical proceedings the oxygen absorbed, but he has discovered a method of directing the boat under water by steam, as if it were on the surface. He has engaged to start from any harbour in France and to reach the coast of England, though navigating under water."

Numerous submarines of various types were built during the following half century by Frenchmen, Americans, Germans, Russians, Italians, and Brazilians. One of them, the *David,* was used during the American Civil War. But the submarine was still in the experimental stage. It became operational, and of vital importance, during World War I, when diesel- and electric-powered German U-boats sank Allied shipping at a rate of more than 500,000 tons a month. Submarine warfare—and antisubmarine warfare—came into their own.

The Treaty of Versailles in 1919 decreed the scrapping of the German U-boat fleet, but the rest of the world now engaged in an underwater-arms race. It became so threatening that in 1921 the five largest marine powers, the United States, Great Britain, France, Italy, and Japan, convened the First International Conference on Limitation of Naval Armament. Like so many of its successors, the Conference did not limit anything. The race escalated, with improvements in submarines engendering improvements in submarine detection, such as electronic tracking devices, which in turn stimulated further improvements in submarine construction.

In the 1930s the Soviet Union developed its submarine fleet, and Nazi Germany, making up for lost time, produced a series of big, fast, deep-diving U-boats. The average length of a submarine was now between 275 and 300 feet, with about 1,400 tons surface displacement; it had diesel engines for attaining a speed of 20 knots on the surface, and electric motors for submersed navigation at a speed of 9 knots. Typical armament was ten torpedo tubes, two deck guns, and a number of machine guns. The development of the gyrocompass, underwater communications, under-water photography, and the snorkel-submarine equipped with a "breathing tube" that permitted its diesel engines to work during submersed navigation increased the efficiency of submarine warfare by the time of World War II. And it did not take long, after Hiroshima, to apply atomic power to the propulsion of submarines as well as to weapons. The prototype, *Nautilus,* made its underwater crossing beneath the North Pole in 1958. The U.S. Navy now includes about 100 atom-powered submarines ranging from the *Skipjack* class of small ships (length 252 feet, displacement 2,830 tons) to the *Lafayette* class (length 425 feet, displacement 7,000 tons).

Military submarine construction had a considerable spin-off in the improvement of submersibles for peaceful purposes—the Russian civilian submarines of the Ministry of Fisheries; the American and British submersibles and semisubmersibles (FLIP: Floating Invertible Platforms); oceanographic vessels, some of considerable size, to mention only a few. The larger variety of atomic submarine, of the *Lafayette* class, may well become the passenger ship of the future.

Another recent addition to maritime traffic, the aircraft carrier, has a prehistory that extends almost as far back as the history of the airplane. An American pilot took off from and landed on the cruiser *Birmingham* in 1910. The first real airplane carrier, *Argus,* was British. She was launched in 1918, before the end of World War I. During the last half century the carrier has grown to mammoth size. The *Enterprise* of the United States is among the largest ships ever built. She is 1,102 feet long, displaces 87,000 tons, and is powered by eight atomic reactors. She carries a crew of 4,600 sailors and airmen, and up to 100 planes which can be launched by four steam catapults at a rate of one every fifteen seconds. She is equipped with radarscopes and electronic sensors to gather battle data, and computers to present fluorescent maps and status boards and formulate strategy alternatives in split seconds.

Luxury liners of the *Normandie* or *Queen Mary* type are obsolete as a means of passenger transport, though smaller versions may survive for a time as deluxe hotels cruising in balmy seas. But jet-propelled hovercraft are coming, capable of reaching surface speeds of 100 to 200 knots; unmanned vehicles will traverse the oceans, operated by remote control. Japanese, Russian, and American shipyards are already building atomic-powered giant freighters of the *Savannah* type.

The development of the contemporary giant oil tanker defies the imagination. The Very Large Crude Carrier, or VLCC, and the even larger ULCC, or Ultra Large Crude Carrier, are a new species of ship. Gulf Oil Corporation operates six Japanese-built tankers of the *Universe* class, each with a carrying capacity of 312,000 dead-weight tons, and this size is quite common now. But there is yet more to come. The Japanese have launched the 476,094-deadweight-ton *Globtik Tokyo* for Globtik Tanker Co. of Great Britain, and a 500,000-deadweight-ton tanker is under construction for Stratis Andreadis of Greece. Preliminary contracts have been signed with Globtik Tanker Co. for a 700,000-deadweight-ton tanker. The VLCCs and larger tankers are almost fully automated and need a very small crew, some thirty men who carry walkie-talkies and ride bicycles on the enormous deck, as large as a hundred tennis courts, plainly bored with their cassette television, game rooms, and bowling alleys. There has been some discussion, in the United States and particularly in Norway, about extending shipboard facilities so that more wives and families can be carried aboard, and of making the ship a largely self-governing occupational village.

The difficulties besetting the VLCC are not only social; they are technical as well. The VLCC needs deeper water than is available in most straits; it may operate at speeds of 15 to 17 knots, but cannot steer at less than four or five; it needs nearly a mile to turn around unaided, and almost four miles and twenty to twenty-five minutes to stop from cruising speed. And this is only if all goes well.

The evolution of these monsters has perhaps already reached its peak. In 1974 the world's then largest tanker, the 484,377-ton *Nissei Maru,* was completed in Japan —and went straight into lay-up. Partly because of shipping overproduction and partly because of a reduction in oil shipments in the wake of rising costs, the tanker market was facing total collapse. Some of the Arab nations, such as Saudi Arabia, Kuwait, and Abu Dhabi, which suffer chronic water shortages, seriously considered putting the otherwise useless tankers to work as floating tanks for water storage. The world's merchant fleet has grown three times in number and twelve times in tonnage since the beginning of the century. In 1970, more than 50,000 steam and motor ships of 100 tons gross or more were registered in the *Lloyd's Register of Shipping.* The annual average growth of tonnage in recent years has exceeded 8 percent. If this trend continues, the total tonnage will have doubled again by 1980.

While passenger service has dramatically decreased (more than 12 million gross tons of ocean-going passenger vessels recorded in 1939 have now dwindled to less than 3 million gross tons), cargo transport is still growing rapidly. In the 1960s, the amount of cargo transported by sea more than doubled, from 1,110 million to 2,280 million metric tons. Tanker tonnage accounts for almost one-half of the world fleet, but by spending less time in port than other ships, they carry more than two-thirds of world trade.

Ship launchings reached record figures in 1973. Japan is the leader in ship construction by a huge margin, accounting for an incredible 49 percent of the world's total production. Sweden is next, followed by West Germany and Spain. The total tonnage launched throughout the world in 1973 was 31,520,373 tons.

The risk of accidents from navigational error grows as the amount of marine activity increases. Clearly, maritime traffic of this volume poses problems that have never been faced before.

Living off the land has been compared
to a man living off his capital;
living off the sea has been compared
to a man living off his dividends.
This applies equally to energy production.

Roger Charlier,
"Harnessing the Energies
of the Ocean"

CHAPTER SIX

THE OILMEN AND THE ENERGY ENGINEERS

FISHERMAN AND SAILOR ARE TRADITIONAL ACTORS on the ocean scene, but the technological revolution has transformed their roles. At the same time, entirely new actors have come on stage. These are not even ocean people; they are land people, the oilmen and miners. They do not look at the oceans in the same way sailors do, as a highway between countries and continents that belongs to all and no one—"God's road," as Ivan the Terrible called it. Nor do they see it as a commons where fishermen harvest a free and inexhaustible resource.

The oilmen and miners perceive the ocean in a somewhat schizoid way. While they depend on its free highways for the transport of their produce, they also see it as a piece of real estate where property claims can be staked out and sovereignties extended, one that can be carved up and crisscrossed with boundaries, as the continents were. The oceans that were all waves and motion, free-flowing and horizontal for the old actors, are rigid and three-dimensional for the new ones. And the resources they seek belong to the earth beneath, finite and nonrenewable.

Precisely because the resources were finite on land, the oilmen and miners ventured into the sea, hesitantly at first. Then, as demand skyrocketed and science and technology advanced much faster than anyone could have imagined, they penetrated deeper and deeper down, farther and farther out.

Oil is formed when plant and animal debris drops to the ocean floor and decays. If there are suitable folds or other kinds of traps, the oil collects, the concentration and collection process taking millions of years. It also requires a tropical climate to begin with. All oil fields trapped in relatively recent rock formations (up to thirty million years old) are in tropical zones. But with ocean-floor spreading and continental drift, oil fields—and the rocks in which they are embedded—migrate to zones far distant from the tropical zones of their origin, to the Arctic and Antarctic and the North Sea. Rocks of recent carbon dating in areas remote from the tropics are thus not likely to conceal oil deposits. Only old rock—older than 200 million years, whose wanderings over the ocean floor can be calculated from the fossil magnetic record—is likely to be oil rich.

Recent work in the North Sea provides a typical example of the use of magnetic measurement to direct oil exploration. Thomas Gaskell, a geologist for British Petroleum, points out in *Physics of the Earth* that "by measuring the old magnetic-field directions in rocks of different ages found in Britain, it is possible to plot the drift of the Dogger Bank from the equator 400 million years ago to its present position of 55° North. If oil is generally found in rocks formed in the tropics, it is only the North Sea rocks that are more than 200 million years old which are of interest."

Geological conformations that may indicate the presence of oil in the seabed are the so-called salt domes formed by the evaporation of shallow inland seas. They are steep-rising structures that distort the sediment layers above them and thus form cavities in which oil is trapped. Such domes, rising under as much as twelve thousand feet of water and sixteen thousand feet of sediment (in other words, more than five miles down), have been found in the Gulf of Mexico and off the coasts of northern Morocco and southern Portugal, and there may be others in the Bay of Biscay and the western Mediterranean. Those found thus far are all on the continental margin, where our present-day continents meet the ocean. Taking into consideration the continuous process of rifting and drifting of the ocean floor, it is quite possible that oil will eventually be found far out in the deep ocean floor.

Thus the prospector is helped by the pure scientist. Yet prospecting is an expensive and risky undertaking, for the only way to do it is by actually drilling.

An oil well is a steel-encased hole. It is drilled by using a series of bits that cut into the rock. When drilling reaches a certain depth, the bit and drill are withdrawn and pipe is lowered into the hole and cemented into place. This casing protects the groundwater from oil pollution and supports the equipment that controls the well. The drilling then goes deeper, and other, smaller casings are cemented into place, until oil is eventually struck.

The basics of oil drilling are the same on the seabed as on land, but special skills are required to perform under hundreds of feet of water. And the farther out and the deeper down one gets, the more costly the operation.

Offshore oil was first produced in 1894 in California, from wells drilled from wooden wharves. The difficulties were considerable, however, and offshore drilling could not compete with the booming production of oil on land. Not much progress was made during the next forty years.

In 1936, operations began in the Gulf of Mexico, culminating in 1938 in the discovery of the Creole field. In 1948, the first offshore platform beyond the sight of land was completed off the coast of Louisiana in the Gulf of Mexico, and the first offshore pipeline was also completed. The platform, built in fifty feet of water and designed to house a crew of fifty men, is still in operation.

From that moment, the race was on. Over the next twenty years, hydrocarbon exploration was extended to offshore regions of seventy countries. And while it took oil technologists nearly twenty years to penetrate from a 50-foot level of water to a 600-foot level, developments moved swiftly in the next decade after that. Drilling in 1,200 feet of water was achieved by 1967, and it has been predicted that there will be drilling at 6,000 feet before 1980.

As the drills went ever deeper, the old-fashioned fixed platforms became impractical and mobile platforms were developed. The so-called jack-up mobile platform is very successful because it can be mass-produced and towed to any site. It consists of a buoyant hull and legs that can be raised for towing and lowered when on site, so that the platform is jacked up above the waves. Such a platform can be used for depths to 600 feet; for deeper water, semisubmersible platforms are used. These are anchored over the drilling site and stabilized by pontoons lying thirty or forty feet below the surface.

The latest development is the drill ship of the *Glomar Challenger* type. Its position over the drill hole is maintained by "dynamic positioning"—that is, a sonar beacon lies on the ocean floor, and four hydrophones on the ship's bottom home in on this beacon, feeding information into a computer which automatically controls any movement of the ship over the beacon.

In 1974 there were over 200 active platform rigs drilling for oil and gas. Most of these were operating in the Gulf of Mexico and in Venezuela's Lake Maracaibo, but there were also some in Europe, the Middle East, the Caribbean,

Africa, Australia, and Canada. Japan has signed a contract for a self-propelled rig with a platform measuring 391 by 262 feet, able to drill in water up to 1,500 feet, and, with some modification, up to 3,500 feet.

Even more sophisticated systems are in the offing. For drilling at any depth, platforms may be placed directly on the ocean floor. Submersed platforms or "subsea completion systems" will be the final answer. Small submersibles for working crews, watertight undersea coupling departments for the linkage of pipelines, and remote sensors to monitor the flow of oil and shut it off in the event of danger, will complete the system. Sheltered from the turbulence caused by wind and waves, such systems will soon be far less costly than surface or semisubmerged equipment. After a certain point, as John Craven, the architect of the *Poseidon* submarine, put it, "deeper is cheaper." The comparatively higher costs of offshore oil are mainly for development and handling. Geological and geophysical costs are much lower in the sea than on land. Some of the lowest-cost reserves are found, and according to present forecasts will continue to be found, offshore; and progress in research and development as well as engineering will effect further economies in the development, production, and handling of oil in the ocean.

Submerged drilling platforms may be joined with underwater storage tanks of a million-ton capacity. A tank of this type already exists in the Persian Gulf. And a million-barrel concrete storage tank is being built for placement on the seabed in the Ekofisk field in Norway. Underwater storage tanks, which may be bell-shaped and open at the bottom, are built on shore like ships and are then towed to the site. From the storage tank, the oil is loaded on large tankers, or it is distributed through pipelines. The amount of pipeline is constantly expanding—in 1974 alone, over 23,000 miles of pipe were laid. And it is being laid over deeper and deeper ocean floor.

In 1958, 440 million metric tons of oil were transported across the oceans. In 1972, the figure had risen to 1,537 million tons and the world's tanker fleet consisted of 3,359 ships of 10,000 tons deadweight or more. By 1975, more than 60 percent of the tankers were larger than 200,000 deadweight tons.

Why this gigantism?

The closing of the Suez Canal in 1967 originally prompted the construction of the Very Large Crude Carriers, because shipping around the Cape of Good Hope was cheaper in these huge vessels. Now the VLCC and ULCC (tankers rising to or even above 450,000 tons) are in use, and the longer trip on the larger vessels has turned out to be cheaper not only than more long trips in a number of smaller vessels but even than the shorter trip through the Canal in smaller vessels. So there will be no going back to the pre-1967 Canal traffic.

There are now over 200 tanker ports in the world, but only forty-seven of them—mostly in Europe and Japan—can handle ships of this type. Elsewhere the giants have to be accommodated in floating superports on artificial islands. Many more of these will probably be built—off the coasts of the United States, for instance—for the transshipment of oil, either to smaller tankers or to pipelines, for the final stretch of the trip.

According to the National Petroleum Council, 29 million barrels of oil per day are now shipped by tanker, and by 1985 this figure is expected to rise to 61 million barrels per day!

We will consider the impact of this kind of production and this kind of traffic on the ocean environment in a later chapter. We have seen the interdependence between the world's energy system, which today is still oil-based, and the world ocean system, which is both a source of oil and a medium for transporting it. But the oceans and energy are interlinked in a number of other ways, and the importance of the world ocean in the world energy system is rapidly increasing.

Not much need be said about the offshore production and transoceanic shipment of natural gas. The problems are similar to those encountered in connection with oil, except that gas does not pollute the water as oil does when spilled. Gas tanker traffic is growing spectacularly, even more so than oil tanker traffic. The first two Liquid Natural Gas Carriers were built as recently as 1964. At present some thirty-seven are on order, and another twenty-five will be needed to meet estimated demand by 1980. About one-tenth of the world's natural gas production comes from offshore.

Oil and gas are nonrenewable resources. As demand keeps skyrocketing, the bottom of the barrel is becoming visible. There is no doubt that eventually we will hit it. When is not sure—perhaps not for another century or two. But we will probably abandon our fossil-fuel-based economy long before that, for it no longer pays to spend trillions of dollars hauling oil across the oceans from one end of the world to the other. Japan's projected oil needs in 1985, for example, would require an unbroken chain of 200,000-ton tankers at intervals of 25 miles from the Persian Gulf to Tokyo Bay; or, one tanker of that size would have to cover a distance equal to 13½ trips to the moon in order to bring oil from sixty different countries to Japan.

An energy revolution is in fact taking place. The conquest of new technologies, the use of resources which are inexhaustible, nonpolluting, and without geopolitical constraints may basically transform our industrial processes and the social and political order to which they are linked. Does this sound utopian? Think how utopian the airplane was just one lifetime ago, and what it has done to our civilization, and how fast!

The energy revolution is ocean-oriented, and the energy engineers are among the new actors in the drama of the oceans.

The generators of atomic energy have already begun to move in. The advantages of placing fission reactors offshore are considerable. The oceans provide an unlimited supply of cooling water. They also provide, at least thus far, a practically unlimited amount of space away from congested coastal areas where more lives and health might be jeopardized by the presence of fission reactors. Whereas a comparable land facility would preempt 500 to 600 acres of shoreline, the area required for offshore units is approximately 100 acres of ocean bottom, just the area occupied by the breakwater. On land, furthermore, reactors are constructed individually and on site, at considerably greater cost than would be incurred for offshore plants, which would be mass-produced and towed to their locations. Land-based reactors, finally, are exposed to the hazards of earthquakes, which may release radioactive plutonium into the atmosphere. Floating reactors are far less exposed to danger from earthquakes. Of course, not all of the dangers and risks besetting the construction and use of atomic fission reactors and fast breeders are eliminated by going to sea. To this we shall return in Chapter 11. Here we are merely introducing the actors.

Offshore Power Systems in Jacksonville, Florida, is planning a first series of offshore nuclear plants for the Public Service Electric and Gas Company of Newark, New Jersey. The plants will be 1½ kilowatt pressurized water reactors mounted on square barges 400 feet long and wide. They will draw about 30 feet of water, and emerge about 200 feet above the waterline. They will be protected from the ocean's furies by semicircular breakwaters built to withstand wind gusts of 300 miles per hour and able to turn back tides 50 feet high. Power will be transmitted to shore via underground cables. The first of these plants is scheduled to begin operation by 1979.

Thermonuclear fusion plants could conceivably be constructed on the same pattern, once scientific feasibility has been established (which may be imminent) and engineering development has caught up (which merely depends on policy decisions with regard to priorities and investments). Floating fusion reactors would have two immense advantages over fission reactors. The first is that the "fuel," deuterium, is available in unlimited quantities in the ocean, whereas uranium, the fission reactor's fuel, is land-mined and nonrenewable. The second is that fusion reactors would not produce radioactive wastes.

The oceans may yet play another role in the production and distribution of energy.

Anyone who has seen and heard the tide racing in on the coast of Normandy or

Brittany knows the power of the sea. Like a herd of stampeding horses arching their crests, a thousand mouths foaming and a thousand hooves thundering, the tidal wave rolls in at a speed of 90 kilometers per hour between Brest and Saint-Malo. At equinoctial tides, some 18 million cubic meters of ocean water are discharged per second, both at ebb and at flood tide.

The ancients knew about the power of the tides. The Greeks used the tides of the Ionian Sea; the medieval Anglo-Saxons built tide mills in Wales and in Dover Harbor in about 1100. The Dutch used such mills in the Zuider Zee as early as 1200 and introduced them to New York in the seventeenth century. The oldest known treatise on the utilization of tides is by an Italian, Mariano, and was published in 1438.

Roger Charlier distinguishes three basic methods by which tidal energy can be used. One is the "float method": the incoming tide is used to raise a floating mass which, as it falls back to its original position, does useful work. The second method, the one most frequently applied in former times, is the use of rotating paddle wheels mounted on a shaft and activated by ebb and flood. The third method, which has now been applied successfully and on a large scale near Saint-Malo, consists of damming-in part of the sea, thus providing a basin which fills with the incoming tide, and then at low tide releasing the water through turbines, either back into the sea or into another basin.

The Saint-Malo plant consists of a dam across the Rance River, four kilometers from the river's mouth. The dam is almost one kilometer long and seals off an upper basin covering 22 square kilometers and holding about 184 million cubic meters of water. The plant is equipped with reversing-blade turbines which generate electricity when driven by the flow of water at ebb and flood tide and pump water when driven by electricity. Half a million kilowatts of power are generated on the incoming tide and when the flow is reversed.

The plant cost $100 million, but in more than six years of operation no repairs have been needed, and operating costs have been less than those of any other power plant in France. The fuel, obviously, is free and inexhaustible, and there are no costs for waste disposal. Energy generation is generally believed to be nonpolluting, but possible ecological and hydrological disturbances are being investigated.

Tidal plants are feasible wherever tidal ranges are in excess of thirty feet. Such conditions exist in more than twenty locations in France; in the British Channel Islands; in Brazil, Argentina, Australia, India, Korea, Canada; and the Shanghai region in China. Calculations have been made which indicate that a plant built on the Kimberley Coast in western Australia would produce over three million kilowatts, or about fifty times the present production of electricity in Australia! The plant has not yet been built, because there is not yet a sufficient market in Southeast Asia for such a quantity of energy. The Russians have an experimental plant operating near Murmansk and are planning tidal dams for the White Sea.

In North America, the Bay of Fundy offers excellent possibilities. Tides here reach a range of 16 meters, the largest in the world. Various schemes have been under consideration for the last fifty years, including President Franklin D. Roosevelt's plan to construct a tidal power plant on Passamaquoddy Bay which would produce 1,250,000 kilowatts—nearly three times as much as the French plant. But nothing has been done during the last ten years to advance the project. Charlier appraises the potential of the Passamaquoddy plant as follows:

> Today's tidal projects are not any longer those envisioned by . . . Franklin Delano Roosevelt. The new Passamaquoddy project would provide large benefits to the United States and Canada. Low-cost power would finally become available to New England and support development of New Brunswick. Flood damage caused by the St. John River would be checked in Maine and New Brunswick while navigation would be improved in both Passamaquoddy and Cobscook bays. Tourism would provide an increment of revenue; the dams would make a highway link possible between Maine and New Brunswick, and the experience gained would be invaluable to develop power plants in the Bay of Fundy.
>
> Finally, expendable resources should be conserved, while renewable resources should be put to use. Energy from the tides is an eternally renewable resource: it should have been placed, a long time ago, at the service of Americans and Canadians.

Where the tides are lazy, the oceans provide yet other forms of energy. An Algerian expert, R. Dhaille, built an experimental plant near Algiers to harness the power of the waves at the shoreline. In California, Scripps Oceanographic Institution scientists worked with buoys which produced their own electricity on a small scale. They operated on a very simple and inexpensive principle: one buoy was left floating and bobbed up and down; the other was anchored, and the intermittent water flow drove a small turbine-generator. The Scripps scientists are continuing their research with various types of wave pumps. They have calculated that, by summing an average of surf conditions over all of the world's coastlines, 2.5×10^{12} watts of power could be produced. This amount is roughly one tenth of projections of the global power requirement for the year 2000!

On the Atlantic coast of Scotland, a large-scale experiment is in progress. There

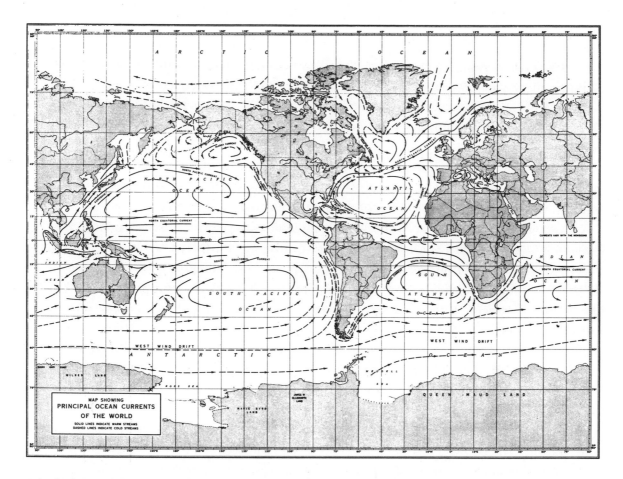

Principal Ocean Currents of the World. Compiled by the U.S. Coast and Geodetic Survey

is a stretch of several hundred miles where the water energy, averaged over the whole year, day and night, is between 50 and 77 kilowatts per meter of shoreline. If only half of this energy could be exploited, it would supply over half the energy needs of Britain. But the engineering difficulties are considerable, and a number of problems must still be solved.

The world ocean is traversed by a system of immense rivers. Driven by winds, listing to the earth's rotation, acted on by chemical properties of the water at particular times and places, these rivers, meandering without riverbeds or fixed boundaries, carry their water masses around the world: the Gulf Stream in the North Atlantic; the Japan Current in the North Pacific; the Brazil Current in the South Atlantic; the Agulhas in the Indian Ocean, and the East Australia Current in the South Pacific, for example. *Panta rhei*—everything flows—and an object thrown into the Antarctic may circle awhile in the south Atlantic, be caught up by the South Equatorial Current, float into the Gulf Stream, pick up the North Atlantic Drift, and end up at the North Pole.

The Gulf Stream carries about 30 million cubic meters of water per second past Miami—more than five times the total flow of all the freshwater rivers of the world. While its surface velocity sometimes exceeds 2.5 meters per second, the average velocity from top to bottom and across the Florida Straits is about 0.9 meters per second.

Thus the total power available is obviously huge. But it is diffused. What can be extracted per meter of vertical cross-section of the stream is small because the velocity is low when measured in conventional hydroelectric terms.

A group of scientists from the Woods Hole Oceanographic Institution and the National Oceanographic and Atmospheric Administration (NOAA) has recently proposed a "honeycomb of turbines" occupying a layer of water between 30 and 130 meters below the surface and stretching about 20 kilometers across the Miami Terrace. "This array of machines," they state ("The Florida Current as a Potential Source of Usable Energy," by Von Arx, Stewart, and Apel), "would deliver perhaps 1,000 million watts on a 24-hour basis"—as much as two large nuclear plants, in fact. The total energy of motion of the Florida Current is equivalent to perhaps 25,000 megawatts, so that the turbine array would extract perhaps 4 percent of the kinetic energy flowing by.

The turbine may not, however, be the most cost-effective solution to the problem of extracting the colossal but diffused power from the oceans' rivers. Another solution is Gerald Steelman's Water Low Velocity Energy Converter (WLVEC). It consists of a horizontal wheel placed below an anchored vessel and attached by a shaft to an electricity generator on the ship. The wheel is activated by a drive cable that forms a continuous loop which has a number of parachute-like sails spaced along its circumference. These sails expand when they face an oncoming current and collapse when they face away from the current.

According to the inventor, the great mechanical advantage that WLVEC holds over turbine-type devices, when operated in low-velocity streams, lies in how much water each can engage at any given moment. Steelman has built and patented a small experimental model, but, as he points out, there are no real limitations on how large a WLVEC could be built. An optimum size might include "sail canopies of 100 yards in diameter and a working loop of 10 miles. A WLVEC built to these dimensions would engage a working (moving) column of water weighing 238 billion pounds or 119 million tons. By comparison, a Kaplan turbine 100 feet in diameter and having a blade strength of 50 feet would engage a working column of water weighing 25 million pounds or 12,500 tons." Steelman estimates that after a few years of technological growth, WLVEC-produced electricity will be the cheapest on the market.

A Bolivian fisherman mends his net

Net fishing and spear fishing in the Arctic. Spear fishing is one of the oldest known crafts. In climatic extremes it continues to be an important tool for survival

A Peruvian anchoveta catch in a traditional purse seine

Opposite page: A dying sperm whale bloodies the waters near
a whaling vessel in the Indian Ocean. Inset: A sperm whale on
the ramp of a waterside processing plant. Above: The initial
dissection of a 60-foot sperm whale on a misty pier

A bountiful harvest of kelp in Kiangsu Province, China

Chinese fishermen seen in an early nineteenth-century scroll.
Collection Elisabeth Mann Borgese, Santa Barbara, California

The papyrus boat *RA II* crossing the Atlantic in 1970

Chinese junks near Macao

Reed boats are still in use

Schooner off Waikiki, Hawaii

The Carta Pisana. About 1275. The first European sea chart

Matteo Prunes. Spanish chart showing the Mediterranean, North Africa, Western Europe, and the British Isles. 1550. Vellum. The Library of Congress, Washington, D.C. Geography and Map Division

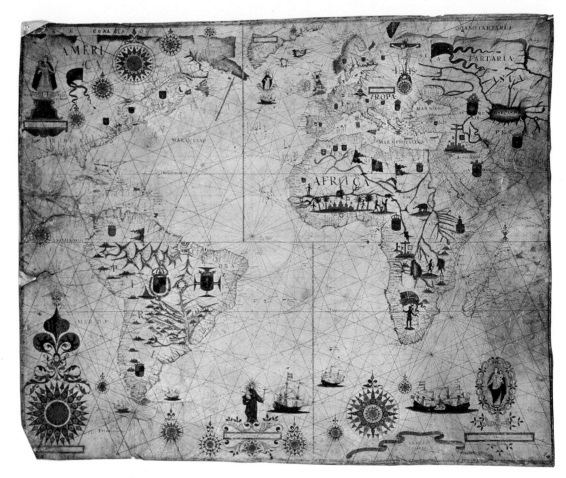

Pascoal Roiz.
Portuguese nautical chart of
the Atlantic Ocean. 1633. Parchment.
The Library of Congress, Washington, D.C.
Geography and Map Division

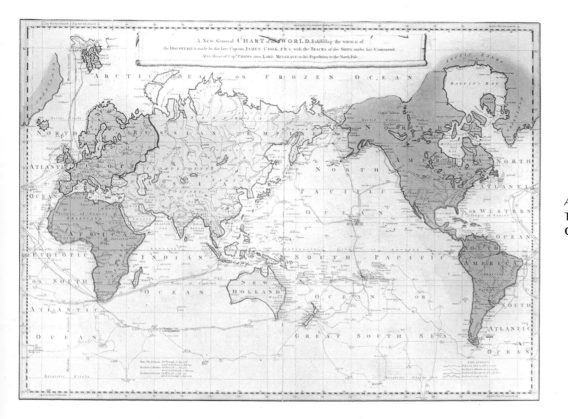

A New General Chart of the World. 1787.
The Library of Congress, Washington, D.C.
Geography and Map Division

Underwater crude-oil storage tank being towed to its location 70 miles off the coast of Dubai in the Persian Gulf

An exposed salt dome in Helgoland, Germany. From T. F. Gaskell, *Physics of the Earth*

Ekofisk facility in the North Sea

Oil refinery in Benicia, California

A test pipeline in Alaska

Genius appears when genius is needed. The so-called energy crisis certainly has acted as a catalyst. Inventions are piling upon inventions. Many will be discarded by history, some rightly, some merely because their time has not yet come.

The undersea vehicles of the future may be driven by hydraulic rather than by electric power, Robert Morgan predicts. Dr. Morgan, Director of the Marine Resource Unit of the Portsmouth College of Technology, proposes a seabed tractor for mineral development; seabed vehicles for catching flat fish and oysters, scallops, and other mollusks; and a combine harvester for the sea, for harvesting seaweed—all driven by ocean currents.

Thus the great ocean rivers may yield kinetic energy. And there is ocean energy in yet another form, and that is thermal energy. Seawater is always cold at the bottom. Cold Antarctic bottom water moves toward the tropics, where it is warmed, and then moves again in tremendous currents toward the poles. Deep seawater may be from 15°C to 25°C (59°F to 77°F) colder than surface water. Taking up and modernizing a scheme first developed in France by George Claude in 1928, two energy technologists, Donald F. Othmer of the Polytechnic Institute of Brooklyn and Oswald A. Roels of the Lamont-Doherty Geological Observatory, point out that the conversion of this heat differential into electrical energy would yield a continuous supply much greater than mankind can now use.

Let us look at the Gulf Stream from the point of view of thermal energy. About 2,200 cubic kilometers of water per day, with a heat differential of 25°C, is at our disposal. If all the warm water were to be discharged to water colder by 25°, this would generate—according to two other workers in the field, James and Hilbert Anderson—182×10^{12} kilowatt hours per year, which is over ninety times the projected demand for electricity in the United States in 1980.

Great temperature differentials between top and bottom layers of ocean water exist in many places—in the South Pacific and South Atlantic, the Caribbean, the Mediterranean—wherever the ocean is at least 1,200 feet deep and has a surface temperature that remains at about 21°C (70°F "Viewing a bathymetric chart," according to William Heronemus, "one can see ocean thermal-difference machines in international waters in the Gulf of Guinea providing all of Europe's energy via hydrogen pipelines"—besides, of course, furnishing the bordering African nations with "an astronomical yield of useful electricity."

Japan, which is so efficient in distant-water fishing and processing, could develop analogous technologies for harnessing the oceans' thermal energy. Heronemus envisions floating tropical cryogenic factories and carriers which would bring the product—energy stored in the form of hydrogen—back to Japan, which could thus in a rather short time make itself independent of the Middle East oil producers.

As for the developing nations, Heronemus points out that "the poorest of nations may have the richest of energy resources lapping their tropical or subtropical beaches: a renewable resource, waiting only for the application of technology and reasonable capital investment. In-country fabrication of ocean thermal-difference power plants appears to be feasible wherever shipbuilding can be accomplished. It may be desirable to purchase turbines and heat exchangers from a more industrialized nation." Or else floating, self-propelled plants, like floating drilling platforms, might be mass-produced and then towed to site.

The simplest of many possible thermal-power generation systems provides for the cooling of warm surface water in an evacuated chamber, thus producing low-pressure steam. The low-pressure steam drives a turbine, cools further, and is finally condensed on tubes through which cold water from the bottom layers is passing and being warmed. The condensate is fresh (distilled) water, and the plant thus produces both energy and fresh water, which brings down the cost of both.

Alemco Inc. in New York has built a substantial plant using low-pressure steam with a condenser for fresh water. A new plan, for a 7,180-kilowatt (net) power plant promises an output of 6 million gallons of fresh water per day at a total installed cost of $18.4 million.

J. Hilbert Anderson and James H. Anderson estimate that sea thermal plants can be built for about $200 per kilowatt, whereas nuclear plants cost $700 per kilowatt ("Sea Thermal Power," by Mark Swann).

Sea thermal plants become yet more economical if the production of energy and fresh water is combined with the production of food through mariculture. This can be achieved by passing the water, before its conversion into low-pressure steam, through shallow ponds where plankton (diatoms) are cultivated on which shrimp and lobsters can feed. The water may be further heated in the shallow ponds, thus raising the temperature differential. Wastes from the shrimp and lobsters can be absorbed by seaweed culture (and the seaweed can then be processed to obtain agar or carrageen) before the water is returned to the plant. Such a complex is now in operation on the coast of the Caribbean island of Saint Croix.

A further, amazingly simple but highly effective improvement to this system has been proposed by a German inventor, Nikolaus Laing. He heats the water with solar-heat collectors of a type he has developed, which occupy only about one-tenth of the surface normally needed for solar-heat collectors.

The artificial island ("Produktions Lagune") he designed would measure one square kilometer and could operate wherever there is an average of 3,000 hours of sunshine per year—which means, in most oceanic regions within the thirty-fifth latitude. This unit would produce the following values per year:

fresh water	4.2×10^6 cubic meters
electricity	3.3×10^8 kwh
animal proteins	2×10^7 kg
fertilizer (excrement)	11×10^6 kg

The total cost of the installation today is $60 million, but will undoubtedly be reduced in the future.

The energy revolution thus brings our new actors back from a land-oriented concept of the oceans as pieces of real estate to a concept more like that of the mariner and the fisherman, of the oceans as oceans, flowing without boundaries as an ecological whole. We are moving from a "territorial" to a "functional" concept, from the consumption of nonrenewable to the use of renewable resources; from "polluting" to "nonpolluting" technologies.

Here we go grubbing about in the earth
for our metals and chemicals, while every
element that exists can be found
in sea water. The ocean, in fact, is a kind
of universal mine which can never
be exhausted. We may plunder the land,
but we'll never empty the sea.

Arthur C. Clarke, *Tales
from the White Hart*

CHAPTER SEVEN
THE MINERAL MINERS

THE SAME TRANSFORMATION IN CONCEPTS, uses, and technologies is in store for
the miners. Like the oilmen, they came from the land and brought with them
their concepts of property and boundaries. They extracted substantial quantities of
coal from offshore, mostly by tunneling from land. They mined sand and gravel that
was needed for land fill and for the production of cement and prestressed concrete.

Some gravels contain diamonds, especially along the coasts of South Africa
and southwestern Africa. Diamond mining there has been in the hands of a
consortium, the De Beers-controlled Marine Diamond Corporation. But Soviet,
Western European, and United States interests have also been active in the area.

After the completion of seismic surveys and seabed samplings by a tug, the
diamond-mining operation is carried out by a specially adapted barge. The barge,
kept steady by six anchors, is equipped with two flexible steel pipes, each up to
sixty feet long. One of them sends water at high pressure to the bottom, to disturb
the sediment; the other sucks up the gravel for sorting and processing on board.
Marine diamonds are not large, but they are of excellent quality.

While much remains to be explored, and there may be diamonds off the coast
of Angola, the Congo, Gabon, Sierra Leone, and Ghana as well as South America
and Asia, the marine diamond-mining industry, which boomed in the 1960s,
has more recently experienced a decline.

Gold and platinum, mercury and chromium have been mined from the oceans. The Australians recently found gold deposits estimated to be worth about $100 million near their coast. Chromite and rutile, barite and ilmenite are extracted from the continental shelf. About 95 percent of the world's reserve of rutile, located on the continental platform of eastern Australia, is mined at the rate of about 450,000 tons a year.

Tin ores have been mined for many years in Malaysia, from a depth of fifty meters. But that was just the beginning. In 1971, an American firm experimented successfully with a hydraulic dredge which worked at a depth of 1,000 meters. The Japanese are doing even better. Their tin-ore dredge scrapes the ocean bottom at depths up to 4,000 meters. Aboard ship the ore is automatically separated from the sand and mud, which are then returned to the ocean.

The Russians are working with ore-prospecting ships specially equipped with a probe containing a radioisotope sensor which indicates the presence of lead, gold, and manganese through gamma-ray absorption. They have reported large tin reserves in the Yakut Autonomous Soviet Socialist Republic and in the Japan Sea. They have also initiated a program for the large-scale extraction of diamonds, platinum, and gold from the Lena River, the Sea of Okhotsk, and along the coasts of the Kamchatka Peninsula.

Mention should also be made of the economic potential of muds which are deposited on the ocean floor as a result of erosion and from alluvial accumulation. There are two kinds of mud, the calcareous and the red. Calcareous muds originate from shell deposits and occupy about 35 percent of the ocean floor at depths anywhere from 700 to 6,000 meters. These sediments are, on the average, 400 meters thick. They contain calcium, potassium, and barite, and can be used for the production of whitewash and fertilizers. Roger Charlier has calculated that this calcareous cover spreads over 128 million square kilometers and accumulates at a rate of one billion tons per year—eight times the quantity yielded by contemporary land operations.

Red muds contain aluminum, iron, copper, nickel, cobalt, and vanadium. They are believed to cover half the floor of the Pacific Ocean and about one-fourth of the Atlantic seabed, at an average depth of 3,000 meters.

In 1970, the total value of worldwide production of mineral resources from the sea was estimated as $1 billion. Of this amount, coal production yielded about $335 million, salt $173 million, sand and gravel $150 million, magnesium metal $75 million, tin $24 million, heavy mineral sands (ilmenite, rutile, etc.) $13 million, diamonds $9 million, and iron sands $3 millon. These figures do not include oil and gas, which produced some $6.1 billion at that time, nor do they take into account two spectacular developments which began in the 1960s.

In 1965 a sensational discovery was made in the middle of the Red Sea, along the prolongation of the line of the mid-ocean ridge which runs through the Indian Ocean. There are pools filled with very hot, very salty water along that line, and at the bottom of these pools rich deposits of metal have accumulated, especially iron, manganese, zinc, and copper. Some of these deposits are 300 feet thick. American oceanographers, working from Woods Hole's research ships *Atlantis II* and *Chain*, estimated that in one pool alone, the so-called Atlantis II Deep, there were sediments containing $1.5 billion in copper, zinc, silver, and gold. A subsequent analysis, also conducted by Woods Hole scientists, led to even more optimistic conclusions. According to these calculations, the mineral deposits in the Atlantis II Deep have a value of $2.3 billion, including $780 million in zinc, $1.1 billion in copper, $280 million in silver, and $50 million in gold. Exploration is continuing under leases granted by the government of the Sudan.

The most discussed and dramatic development, however, has been that of manganese-nodule mining from the deep ocean floor of the Pacific.

Nodules were first dredged up by the *Challenger* expedition in 1872, between Honolulu and Tahiti. Where did the nodules come from? Were they of cosmic origin? The *Challenger* scientists wondered if they had entered the sea as incandescent particles thrown off by meteorites in their passage through the atmosphere (Eric Linklater, *The Voyage of the Challenger*).

A hundred years later, we still are not quite sure about the nodules' genesis. We know where they are: the richest deposits occur in a narrow band, perhaps 125 miles across and 1,000 long, running roughly east-west along the southern edge of the equatorial belt at a depth of about 5,000 to 12,000 feet, in the Pacific, Atlantic, and Indian oceans. We know what they contain: iron, nickel, copper, cobalt, and traces of two dozen other metals, in addition to manganese, often in concentrations comparable to those in land ores. Rich beds contain about ten kilos of nodules per square mile of ocean floor. As they vary in composition, so they vary in shape and size. Many look like potatoes, and there are trillions of tons of them scattered on the seabed—1.5 trillion tons in the Pacific alone.

In a way, these nodules are a "renewable resource," for they keep re-forming. Growth rates between 1 and 100 millimeters per million years have been reported. This is slow, but considering the vastness of the resource, it adds up to a great deal. It has been estimated that 16 million tons of nodules are formed every year.

How they form, and how they stay on the ocean floor without being buried by other sediment, is not presently understood, but recent research has given rise

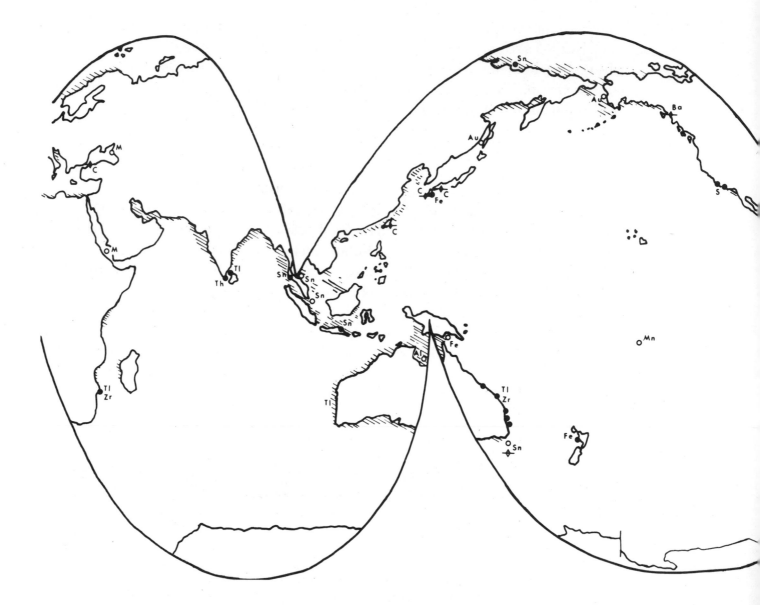

to some fascinating speculations. In part, at least, the nodules may be the accumulated work of living creatures—as coral reefs are. Certainly they are the habitat of protozoans and other microorganisms. J. Greenslate of the Scripps Oceanographic Institution suggests that they are formed around a skeleton of tubules and chambers left by deep-ocean Foraminifera. The manganese content may derive, at least partially, from the activities of bacteria capable of oxidizing and precipitating this metal (we do not know how to do this, but a number of animals do). The other constituents of the nodules—iron oxides, copper, nickel, and other metals—probably precipitate inorganically, but even the copper and nickel may

Crude Oil
Pt Platinum
K Potash
Al Bauxite
Au Gold
Ba Barite
Fe Magnetite
Cr Chromite
Mn Manganese nodules
C Coal
D Diamonds
M Metaliferrous mud
Ca Calc. sand, shells, aragonite
S Sulfur
Sn Tin
Th Monazite
Tl Ilmenite, Rutile
Zr Zircon

Unconsolidated deposits
● operating
○ developmental

Consolidated deposits
◆ operating
◈ developmental

Composite map based on research by Dr. R.H. Charlier and maps which have appeared in *Ocean Industry, Oceanology International, Undersea Technology,* and a petroleum occurrence map by Professor K.O. Emery of Woods Hole, Massachusetts. Drawn by Anita Brandes

have been concentrated from seawater by plankton and deposited in the sediment.

The study of the North Pacific nodule beds started very recently. It is to be hoped that human greed will not bring it to a premature end. For there is wealth in the nodules, besides unique scientific interest, and the technologies for scooping them up by the millions of tons a year and for processing their valuable metal content are developing much more rapidly than the leisurely speculations about their genesis and interaction with the oceans' living forms—and therefore, perhaps, their renewability.

The splendidly equipped German research ship *Valdivia* has been exploring the nodules of the North Pacific for the last five years. German companies are cooperating with French, Japanese, United States, British, Soviet, Austrian, Canadian, Dutch, and Australian ones. But in 1974 Howard Hughes stole the show and the headlines.

There are two basic methods by which the nodules can be dredged up. One is the so-called continuous line bucket (CLB) developed by the Japanese. It consists of a long loop of cable to which buckets are attached at intervals. A traction drive moves the cable so that the buckets dive into the deep, drag across the seabed to scoop up nodules, and come up again to empty their load. Such a loop can be attached to any ship, and sometimes two ships are used, to keep very long cables from tangling.

The second method, developed by Deepsea Ventures but perfected by the Hughes system, is hydraulic. The nodules are sucked up through a long, flexible steel pipe—somewhat similar, in principle, to the marine diamond-mining system.

Hughes's nodule-mining ship, the *Glomar Explorer,* was built in great secrecy. Its most spectacular new feature is a roofed 300-by-100-foot submersible mining barge, a kind of submerged sucking (rather than drilling) platform, which is held in place below the bottom of the *Glomar Explorer* by 150-foot docking legs and tubular stability controls. A steel pipe passed through the ship's bottom is connected by divers to a dredge head, 50 feet in diameter, on the barge. Then the vessel lowers a 4,000-ton dredge pipe thousands of feet long. As the ship moves slowly over the ocean, the self-propelled bottom equipment sucks up nodules like a giant vacuum cleaner.

Just when prices of land-mined minerals are beginning to soar, this sea-based operation, at the moment still free of concessions, license fees, and taxes, is expected to make huge profits. Compared with mining on land, where $9 of capital may produce $1 a year, $1 of capital will return at least $3 from ocean mining.

John Mero, a pioneer in the nodule mining industry—only five years ago, most people smiled at him as a foolish dreamer—predicts with the authority of one whose dreams have come elegantly true that massive production of minerals at one-fifth or one-tenth of land prices will signal the end of mining on land and of its vested interests. Not only the ecological, but the economic and political implications will be vast.

The nodule miners are mutants. From land people, they are turning into ocean people. They need no fixed installations, no territorial property. They sweep over the ocean floor, mobile, beyond political boundaries. Whether they will ever realize their full economic potential, as it is now conceived, remains to be seen. Once mining moves fully into the sea and men's minds adjust fully to the ocean's enormous proportions, its eternity, its cyclicity, they may begin to look at the problem of resource production in an entirely new way. For the ocean itself can be looked at as a "liquid mine" containing every conceivable mineral in inconceivable quantities.

Just like the inexhaustible, renewable energy resources of the ocean, so these mineral resources are diffused—so much so that nonrenewable energy resources are not adequate for their recovery. But if we adjust our minds to the concept of inexhaustible energy, then logic brings us to the concept of inexhaustible resources that can be extracted from the waters of the world ocean and recycled on a gigantic scale—resources that keep welling up from the interior of the earth and keep being returned by the rivers and the atmosphere.

The production of inexhaustible, renewable energy from the oceans, whether in the form of thermal or thermonuclear energy, becomes economical when it is combined with the production of fresh water—which is becoming scarce—and the production of food by mariculture. Always assuming the availability of sufficient energy, we may now add the production of minerals, which may be extracted from the concentrated brines left from the desalination process. Irving Kaplan ("Mater Omnium . . .") has designed an integrated factory system combining energy production, mineral production, waste recycling, fresh-water production, and aquaculture.

The main difficulty in applying energy on such a vast scale is heat waste or thermal pollution. Even ocean-thermal energy, although it subtracts heat from the ocean rather than adding it, may raise the temperature of the ocean ever so slightly in the long run. And the production of any form of energy adds heat to the atmosphere, perhaps raising its temperature by one or two degrees in a century. This may not seem much, but even small alterations on a large scale may have irreversible consequences, at least on our human time scale. There are, however, ways to cope with the problem—by reflecting the heat back into outer space, for instance, or by towing icebergs from the Arctic and Antarctic regions to the sites of energy production and using them as coolants. That it is possible to utilize the oceans' currents to transport them has been demonstrated by John Isaacs of Scripps, who has suggested that icebergs could be used for fresh-water production. If, instead, we derive fresh water as a by-product of energy generation, we might use the icebergs as coolants. At any rate, disposal of heat waste must be an important part of this integrated system.

Minerals from the liquid ocean mine might also be extracted by biological systems; that is, we might use animals to do the job for us. For instance, the ascidian, a sea worm, concentrates vanadium into itself; the lobster extracts copper from sea water; and T. F. Gaskell *(Physics of the Earth)* points out that "many sea animals have perfected ways of concentrating minerals. . . . Research into this problem may soon lead to the devising of man-made equipment for performing the extraction under controlled conditions. . . . By copying the methods of these animals it should be possible to devise adsorption and filtration processes for operation on a large scale." By genetic engineering, James Danielli of Buffalo University has produced a strain of bacteria which, given sufficient sunlight, extracts hydrogen from ocean water, which can then be recovered for use as fuel. The waste of these bacteria can be used as fertilizer, another scarce commodity at a time when energy costs are high.

Integrated systems of the kind we have been discussing would bring together all the actors in the drama of the oceans: the fishermen and mariners, oilmen and energy engineers, mineral miners and scientists whom we have already met, and the architects, shore developers, and warriors, whom we are going to meet next. This is indeed the pervasive nature of the oceans—this ecological whole where each of the multiple uses of space and resources interacts with every other use. It forces a systemic, organic approach to the world's problems.

There still are no architects today.
We all are merely the precursors of him who,
one day, will deserve again the name of architect,
for it has been said: The Master, who builds gardens
out of deserts and piles miracles unto Heaven.

Walter Gropius in 1919, quoted
in *Planet Meer* by Jürgen Claus

CHAPTER EIGHT

THE ARCHITECTS

WHEN YOU LOOK AT A SWISS CHALET, you are not very likely to be reminded
of the ocean. Yet there is an old evolutionary link between the two. If you strip
away the stone walls enclosing the ground floor (often used to house cattle) while
leaving the supporting beams and the wooden top floor (containing the living
quarters and the balcony) what you see is a pile dwelling with its front terrace for
the landing of boats. This kind of pile dwelling is found in many sites, both
primitive and archaeological, from New Guinea to Central Africa, from the Gulf
of Maracaibo to the Swiss lakes.

The pile-dwelling village is one expression, widespread in time and space,
of the interaction between man and water in the tidewater zone between land and
sea, where this interaction has always been most intense.

There are other architectural expressions of this interaction. The water
surrounding an island may have a crowding effect, squeezing buildings into high-rises,
like mountains rising when continents clash.

Or the water may be used as an element of communication, with canals taking
the place of streets, as in Venice, Bangkok, and ancient Amsterdam. The relative
quiet of waterborne traffic, with no clatter of wheels or hoofs on hard pavements; the
bobbing and splashing of floating markets; the gracefulness of social functions

flowing along on the water—weddings on barges, and funerals; the facades of houses mirrored on the water street—all these give the water city a unique charm and constitute a unique challenge to urban designers. And now that modern technology has given the designers tools to create artificial islands, and to build down from the water surface as well as up, a new age of marine cities may be dawning, one which may even cure some of the urban blight that epitomizes the crisis of our land-oriented civilization.

Marine architects can be divided roughly into two groups, although these overlap. The first consists mainly of industrial architects. Their basic purpose is to relieve coastal congestion by moving especially polluting industries out into the ocean, to an acceptable distance from human settlement. Their approach is essentially that of the land reclaimer; their artificial islands are *lands,* rooted in the continental shelf and built of sand, gravel, and rock. What they have in mind, basically, is to extend the existing industrial system from land without transforming it.

The second group has a fundamentally different aim, and that is to take human society, including its industrial system, out into the new environment, of the oceans, thereby transforming it. These architects think in terms of floating cities, of new materials, of integrated processes, of ecological wholeness, of a new type of human society.

The first group is exemplified by the Bos Kalis Westminster Dredging Group, which has elaborated plans for three types of artificial islands in the North Sea, on the borderline between the Dutch and British continental shelves. These plans include:
—small islands, of about 50 hectares, for highly specialized uses such as centralized waste treatment, central storage, the processing of gas and oil extracted in the vicinity of the island, and the generation of energy together with a fresh-water plant. The small islands can also be used as first building blocks for larger ones;
—islands with an area of about 300 hectares, which would contain air terminals, specialized industries, and harbors where emergency repairs could be undertaken;
—very large islands, with an area of 1,000 hectares or more, which would be used for extensive deepwater-bound industries, including power plants and fresh-water plants.

Comprehensive studies have been made by the Group to determine the optimum locations for such islands and the availability of building materials—sand, gravel, rock.

The planners reached the conclusion that "a small, relatively costly island, specially arranged for the processing of waste matters, can be economically justified. When proceeding from the thesis that industrial growth is indispensable for the maintenance of the welfare level, it is certain that financial sacrifices will have to be made regarding environmentally burdening activities, and that the creation of industrial sites outside population concentrations will in the near future become a necessity. Investigations will have to be carried out as to how larger islands can be capable of offering solace to industries that are searching for development space" *(Sea Island Project . . .)*.

Another example is Ocean Industries' offshore aragonite mining project off Bimini in the Bahamas. Their 200-acre artificial island, named Ocean Cay, was formed by dredging sand from the sea floor. Aragonite is a natural form of limestone precipitated to the seabed in the form of calcareous sand. It is used in the manufacture of cement, chemical lime, steel, glass, pulp and paper, agricultural fertilizers, and as a source of carbon dioxide for carbonated soft drinks. It is suction-dredged, dehydrated, and stockpiled on Ocean Cay.

No one lives on Ocean Cay, but there has been fishing in the zone, and the question of eventual conflict between dredging and fishing has not yet been resolved. There is continuous monitoring of the operation, and a study is being made to determine whether and where "buffer zones" prohibiting dredging should be established.

A third example is an artificial-island type of terminal recently constructed eight miles off the coast of Brazil, near Areia Branca in the state of Rio Grande do Norte. It is intended for the transshipment of solar salt (salt from brine evaporated by the sun) and possibly other bulk commodities. The island, Termisa, is considered a milestone in the history of offshore terminals. Twenty-four circular cells of sheet-piling, lowered to the sea bottom over guide piles and driven about fifteen feet into the ocean floor, form the perimeter of the island. Each cell is filled with material dredged from the seabed some nine miles from the site. The island includes a barge-unloading wharf, a storage area, and auxiliary facilities. It is uninhabited.

Even though these islands are conceived by specialists for specialized purposes, the compactness imposed by the surrounding sea generates a certain integrative approach that may have been alien to the original intentions of the marine architect. On the smallest of these man-made islands we see an integration of waste disposal, energy generation, and production of fresh water that transforms the individual processes into a much cleaner, more economic system. This is where the future begins.

No new technological breakthroughs are needed to progress from Ocean Cay to such marine cities as Ocean Venice, which was designed by a team of Britain's leading architects and engineers for the Pilkington Glass Age Development Committee. The plans were unveiled in London in 1971.

The city, to be built on concrete stilts and protected by a floating breakwater and curved 180-foot-high city wall, is designed for a community of 30,000 inhabitants. It could be constructed in any of the shoal waters that exist on over 10 percent of the surface of the ocean: in the Yellow Sea and the East China Sea; off the coast of Israel and in the Persian Gulf; off the South American coast and in the Gulf of Mexico; in the Java Sea and in the Gulf of Siam; in the Baltic, the Adriatic, the Black Sea, the North Sea, or the Irish Sea.

The site selected for Ocean Venice is in a shoal area fifteen miles off the east coast of England, where the water is less than thirty feet deep and the tidal range is no more than four to seven feet.

Power installations will be located in the terraced city wall, which will surround the marine city like the rising tiers of a huge amphitheater—a modern reconception of the city walls of Dubrovnik. These power installations are to be fueled by undersea natural gas, deposits of which have been discovered nearby. Indeed, the discovery of offshore natural gas has made the concept of building cities offshore very much more attractive. The upper tiers of the terraced city wall will contain a great number of apartments, connected by escalators, travelators, and covered sidewalks with one another and with the shops, restaurants, clubs, and gardens within the city wall. Other, individual housing units, as well as public buildings, schools, offices, and hospitals, will be constructed on islands of floating pontoons held in position by anchor chains. These can be moved around and rearranged in accordance with changing needs, making the urban plan flexible within the fixed limits of the city walls. Transportation within the city will be by electrically powered boats and water-buses; all public buildings will have their own landings, and private boats can be parked alongside the floating islands. Footpaths and bridges will carry pedestrians from island to island through the city. The traffic thus may very much resemble that of Venice, but a Venice modernized and streamlined. Traffic with the mainland will be handled by hovercaft and helibus.

Besides producing energy, the city will earn its keep with a sea-fish farm using temperature-controlled tanks, fish canneries, the manufacture of fertilizer from seaweed, and a desalination plant producing enough fresh water not only for the city but also for export to the mainland. Most food, including dairy products and meat, will be imported from the mainland in container barges.

"A permanent offshore center of this sort is regarded as important to the development of specialized marine industries in various parts of the world," Arthur Nettleton points out. "The notion is not just a pipe dream, as the detailed schemes for carrying it out reveal. Although the first of these cities may not be built until the early part of the next century, they *could* be created today with present materials and existing methods" ("Cities in the Sea").

Although Ocean Venice is still land-anchored, fixed in its position, "territorial"—and may even harbor traces of a glorified offshore drilling platform—it *is* ocean-oriented. It belongs largely to the second group of projects, those that attempt to change mankind by changing his environment. The compactness imposed by the ocean environment; the functionality of space and enclosure; the integration of functions; the integration into nature—all these are features the offshore city has in common both with the land city in its glorious periods of centuries gone by, and with the true marine city of the second group of architects.

One should include in this group of in-between projects Paul Maymont's and Kenzo Tange's plans for the extension of Tokyo into Tokyo Bay. Maymont's bay city, planned to accommodate 10 million inhabitants, is to rest on prefabricated cement caissons, each several hundred meters across. Vertical cities are to be erected on these, with a density of about 2,500 inhabitants per hectare. Each city is shaped like an amphitheater with its bottom underwater (Jürgen Claus, *Planet Meer*).

Kenzo Tange projects a city of 20 million inhabitants. It rests on trapezelike structures, and everything is so arranged as to facilitate communication between home and work and to make sport and entertainment, health care and education easily accessible.

A number of bridge city projects belong in the same group, for example, James Fitzgibbon's Hudson Bridge City, which would house 100,000 people. And two Italian architects, Mario Perugini and Giovanni de Benedetti, have proposed a ring-shaped bridge city to span the Straits of Messina. This tubular structure would link railroad and auto transportation systems, hotels, apartment houses, offices, and public buildings for 400,000 inhabitants.

"Now is the hour of the genesis of floating architecture," Jürgen Claus said (*Pacem in Maribus Proceedings*, IV). "The invention of marine architecture reveals man as the 'creator.' What is called the inner space of the planet earth corresponds in its development to the inner space of man: to his imagination, to his creative mind, to his strong and vital desire not to let himself fall back behind his cultural achievements of centuries, but to accept his own history as competition for his future."

The technological breakthrough has already happened: dynamic positioning, stabilization, and module construction, plus a bewildering array of new materials are its technical ingredients. When positioned, the floating city feels so stable that even in a storm one feels one is on land; yet this city is capable of traveling slowly over the ocean. And it can be added to or reduced as new needs arise. Thus it is an ideally flexible urban structure.

"I have learned," says Buckminster Fuller (*The American Way*, April 1973), "with geodesics, for instance, how to enclose a given clear-span of space of a given size—proof against earthquakes, snow overloads or hurricanes—and to do so with only 3 percent of the weight of materials required by any other known engineering-strategy alternative to that of my triangulated compound structures. I find that I can give you thirty new disaster-proof buildings for one old building of equal capability."

This is clearly of basic importance for a city moving in what used to be considered a hostile environment of waves, storms, and seaquakes, and whose model is the spatial economy of the ship rather than the unboundedness of the terrestrial habitat.

Manganese, which may become available in abundance if nodule mining gets underway on a large scale, may substitute in alloys for steel in the construction of marine cities. It has many advantages, among them resilience and flexibility. Cities might conceivably be constructed of water itself—of ice, as Nigel Calder seriously suggests in *The Environment Game*. Ice would be unsinkable, for one thing, thus adding to the city's safety. This is by no means science fiction. Igloos are not, and, as Calder points out, there was a project during World War II to develop an unsinkable aircraft carrier, made of ice and reinforced by wood, to be kept at a temperature of −15°C by an ice machine. The model, named *Habakuk,* was calculated to withstand navigation in any climate.

Whatever the material of the sea-level platform, one can build up and down from it, up into the sky and down into the sunlit upper layers of the ocean, where life floats by and just looking out the window makes you feel the human drama embedded in, and at one with, the drama of nature in the oceans.

One of Buckminster Fuller's designs, a floating city planned for a million inhabitants, is a tetrahedron, a huge pyramidal structure with each edge about two miles long. "The whole city," Fuller writes ("Floating Cities," *World,* December 1972), "can be floated out into the ocean to any point and anchored. The depth of its foundation will go below the turbulence level of the sea so that this floating tetrahedronal island will be, in effect, a floating triangular atoll. Its two-mile-long 'boat' foundation will constitute landing strips for jet airplanes. Its interior two-mile harbor will provide refuge for the largest and smallest ocean vessels."

In the plans, spacious family living units are located on the outside surface. Recreational activities are organized in the upper layers and in the surrounding ocean. Businesses and large arenas, conference and entertainment halls are concentrated in the central city, while below, in massive columns jutting down into the ocean, the heavy industries and storage facilities are housed.

The city would generate its own energy from the ocean in thermal power plants while using direct solar energy for domestic heating and cooling; eventually it might have a nuclear fusion plant fueled with deuterium from the sea. This energy could be converted to electricity for powering the city internally, or to hydrogen for fueling external transportation to and from the city or for moving the city itself.

The city would integrate its energy production with desalination, mineral production, waste recycling, and mariculture, and thus enjoy a high degree of self-sufficiency that would also enhance its mobility.

An alternative to this tetrahedronal design is Fuller's Triton City, a series of small modular communities which would link together to form a city of from 90,000 to 125,000 inhabitants. The basic unit of this city structure would have 3,500 to 6,500 persons and support an elementary school, a supermarket, and a number of smaller stores and businesses. The city could expand gradually, or be set in place one module at a time. At its full size, Triton City would include a module for government offices, medical facilities, a cultural center, shopping centers, sports grounds, and all the other amenities a metropolis should offer.

In Hawaii, John Craven is directing a multidisciplinary group of engineers, architects, and scientists in developing a floating city that should become operational in the next few years as a marine exposition center off the coast of Oahu. In their systems-analytical approach to urban planning, the members of this group posit fifteen basic requirements for the city (Craven *et al.,* "The City and the Sea"):

1. Economic surplus of food, clothing, and shelter for all the inhabitants.

2. Nonpolluting, rapid, comfortable, flexible, and immediate transportation for people carrying approximately forty to eighty pounds of goods.

3. Nonpolluting and nonobstructive transportation for goods in excess of eighty pounds.

4. Telephone, television, mail, accounting, and computational communications.

5. Nonpolluting, efficient, and effortless disposal of wastes—human, solid, liquid, organic, inorganic, toxic, and nontoxic.

6. Opportunities for social affiliation in recreation, organization, sports, spectator sports, conversation, theater, arts, and music.

7. Opportunities for achievement in organizations, business, intellectual and physical activities, culture, and arts.

8. Opportunities to exercise leadership with family or selected associates, or with pets and hobbies.

9. Opportunities to engage in dogma and independent belief with family or selected associates or through church or religious activities.

10. Opportunities to derive adequate economic income.

11. Opportunities for privacy, individually, in pairs, or in small groups.

12. Opportunities for experiencing spacious and unscarred natural landscapes and environments.

13. Variety and freedom to select or to alter one's local environment.
14. Freedom from noise, atmospheric, land, and water pollution.
15. Preservation of similar opportunity for posterity.

These general requirements will all be met by the Hawaii Floating City Project, as follows:

Food and clothing arrive by container barge, which represents the lowest transport cost now known. Costs for shelter are related to cost for platform, and should be less than mainland costs for a comparable city.

Within the city, transportation of people is by escalator, elevator, and moving sidewalk; forty- to eighty-pound loads are carried by battery-powered shopping carts. The overall dimensions of the city are approximately two square miles by 200 feet high, making any point accessible from any other with a maximum walk of thirty minutes. The average length of time required to reach shopping areas is less than ten minutes; work areas, ten minutes; entertainment areas, fifteen minutes.

For goods in excess of eighty pounds, half- or quarter-container-size freight elevators throughout the complex connect with platforms beneath the lowest deck of the city. There, goods are loaded on barges and taken to other points in the complex or to the main island. Mechanisms for making the barges securely fast to the platform must be worked out, to reduce the hazard of spilling or damaging the contents during cargo-transfer operations.

Communication is excellent, since line-of-sight, high-information-rate micro-wave is always available to the main island or to a satellite, as is line-of-sight, extremely-high-information-rate laser operation. The city's compact size permits low-cost cable connections to each habitation and facility.

Waste management is relatively easy in such a complex. There are direct vertical shafts for solid wastes and pipes for liquid wastes, and containers beneath the lower deck of the platform. If necessary, wastes can be collected in barges for transport to processing and recycling sites. Such recycling can include producing fertilizer for the mainland and nutrients from biologic wastes for fish farming.

Sailing and boating and gymnasiums for other sports are easily accessible in the complex. Transportation to beach parks by small boat and hydrofoil make field sports and beach recreation accessible within ten or fifteen minutes. Theaters and concert halls are within fifteen minutes walking distance. Of particular importance and currently denied to most urban dwellers are plazas, cafés, and socializing areas within less than five minutes walk from each dwelling.

The ease with which people can get together encourages participation in artistic, cultural, and professional activities. Similarly, opportunities for religious assembly of a wide variety of faiths and beliefs are afforded by the proximity of people to each other.

Opportunities to derive adequate income are afforded by a diversity of businesses and manufacturing plants that make use of the extensive spaces available in the submerged columns and the low cost of bringing raw materials to the site. Economies made possible by the floating city's small scale thereby outweigh the inefficiencies of partial isolation.

Opportunities for privacy are afforded by the design and location of dwellings. Parallel placement of units for visual privacy is combined with soundproofing for auditory privacy. Setbacks permit the introduction of sunlight for gardens and open private spaces. At the periphery, density ratios equivalent to ten to fifteen per acre are obtainable, or more than a quarter acre for a family of four.

Since the view from each habitation is outward, an unobstructed view of unspoiled seascape is available. The movement of the platforms results in changes of vista and diversity of views on a scheduled or quasi-random basis.

The privacy of the habitation and its size permits individual adaptation of the local environment to meet individual needs. Perhaps some limitation on generation of odors will be found necessary, but the closed-environmental conditioning systems for each habitation should make this a minor problem.

The ability of the city to move to new environments in the event of ecological accidents, the elimination of the automobile, the ease of access to goods, services, recreation, and community activities, the ability to collect and process wastes, and so forth, should make the city essentially pollution-free.

Finally, the age-old problem of obsolescence and decay, which has plagued city development since the time of ancient Troy, is resolved by the use of the module concept, which permits replacement as new technology develops. Major portions of the older platforms can be discarded as necessary.

The Japanese architect Kiyonori Kikutake proposes Marine City, which consists of a number of artificial islands merging into an "organic body." Islands built upward and carrying high-rises alternate with islands built downward, submarine cities jutting some thirty meters deep into the ocean. The city's supporting services, as well as a number of apartments, are located there. The bottom of the submarine city rests on spheric pontoons which act as stabilizers, and the top surface is a round steel plate several hundred meters in diameter, somewhat elevated above the sea surface and provided with suitable borders to keep out the waves. These serve as city squares, in which communal activities take place.

Paolo Soleri (*Arcology: The City in the Image of Man*) has developed plans for a series of floating ocean cities. Novanoah I and II may serve as examples.

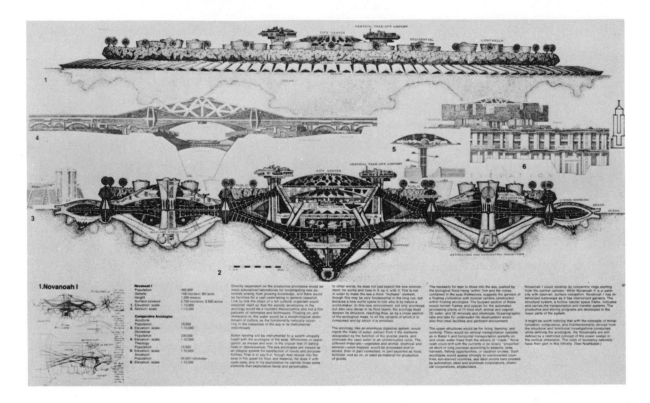

Paolo Soleri, *Novanoah I.*
From Soleri, *Arcology: The City in the Image of Man*

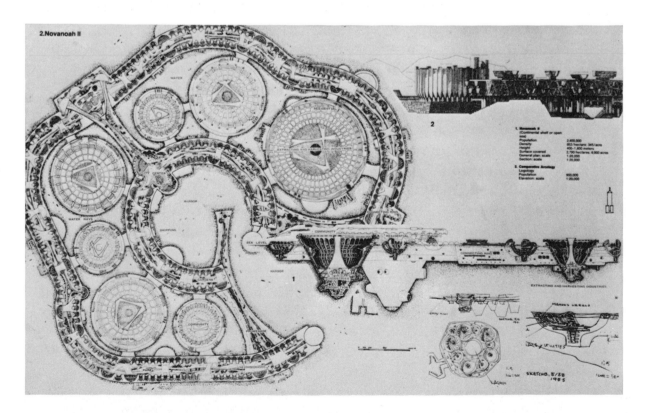

Paolo Soleri. *Novanoah II.*
From Soleri, *Arcology: The City in the Image of Man*

162

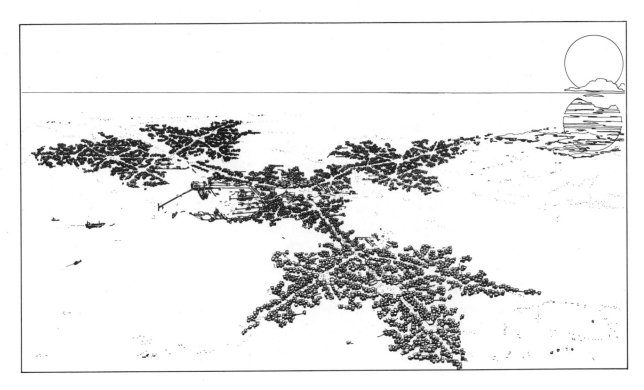

Thalassopolis I. From Rougerie and others, *Thalassopolis*

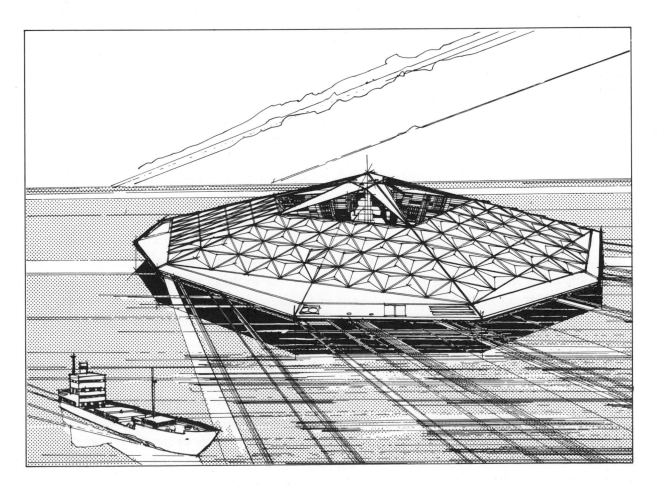

Thalassopolis II. From Rougerie and others, *Thalassopolis*

Novanoah I would accommodate 400,000 inhabitants and cover a surface of 2,750 hectares, or about 6,800 acres. Novanoah II would contain 2,400,000 inhabitants in an area not much larger—2,790 hectares, or 6,900 acres. Novanoah I could develop by concentric rings starting from the central canister. While Novanoah II is planned to be a marine city with open-air surface navigation, Novanoah I would have interior waterways and gardens. According to the plans, the structural system, a hollow tubular space frame, carries the transportation systems. The production and storage programs are developed in the lower parts of the system. Like all floating cities, both Novanoahs are largely self-sufficient economically, generating their own energy from the sea. Like the others, they are mobile, and the whole system can slowly "migrate."

Jacques Rougerie's plan for Thalassopolis I, designed for the zone between Borneo and Sumatra, contains mariculture areas and facilities for international marine research, linked to an international data bank. The city exists on three levels: urban life and civic activities take place on the upper, above-surface level; laboratories and industrial plants are contained in the submarine level; sea farming and connected ecological research are conducted on the ocean floor. All external elements are constructed of cement, to prevent corrosion.

Thalassopolis II, designed for the Sargasso Sea, is really more of a data-gathering platform than a city. Rougerie believes that both his marine cities should be under the jurisdiction of the United Nations and be administered by the appropriate organs of that organization. This raises a number of unsettling but fascinating questions about the whole emerging range of artificial islands and marine cities: What would they do to our land-based nation-state order? Whose laws would they obey? What would be the perquisites of their "sovereignty," were such to be recognized?

Even with fixed artificial islands, problems are bound to arise. Suppose they are located toward the outer end of the ocean area under national jurisdiction, wherever that may be. Would it then be legitimate for a nation to further extend the area under its jurisdiction by surrounding the artificial island with a "territorial sea" of its own? If, on the other hand, the island lies clearly beyond the limits of national jurisdiction, to whom will it belong? Should it be "registered," the way ships are now? Will "flags of convenience" develop for cities as they have for ships, allowing registration with the nation which offers the greatest tax and other benefits?

When we come to floating cities moving around in the oceans, the problems become even greater. One can foresee three developments:

First, the emergence of a new concept of property—that there is no real estate in a city floating in the oceans, and that space can be used, but not owned. The concept of the common heritage of mankind, first applied to the resources of the deep oceans beyond the limits of national jurisdiction, may begin to penetrate civic life as a whole.

Second, it is likely that economic self-sufficiency will strengthen local government and encourage national decentralization. This seems logical. Furthermore, it fits into a general trend of our time.

Third, the mobility of the city, its lack of territorial integrity, will entail a reconceptualization of the territorial state in general, separating the concepts of sovereignty and of territoriality. It makes of the state a functional and cultural entity rather than a territorial one. At the same time it potentially makes of every state a "world state" whose citizens, no matter where they find their island positioned at the moment, maintain their loyalty to the state and participate in its political processes, not because they are territorially beholden to it but because of their cultural affinity.

This reconceptualization also accords with other ongoing trends. In Marxist terms it fits in with the "withering away of the state" and its transformation into a community. More broadly, there is a trend away from territorial and toward functional concepts. The interdependence of all nations, brought about by the technological and other developments of recent decades, is not likely to lead to one territorial world state, but rather to a network of overlapping functional world communities. The floating, "de-territorialized" cities could be an embodiment of this development, with their citizens becoming self-governing in local matters while living under the jurisdiction of an international ocean authority and participating in its decision-making processes.

The problem is man's use not only of the earth
but also the adjoining sea, and the ways
in which he has integrated both into
new sub ecosystems the nature of which is
as yet poorly understood.

Norton Ginsburg, "The Lure of
Tidewater: The Problems of
Interface Between Land and Sea,"
in *Pacem in Maribus Proceedings*

CHAPTER NINE

THE SHORE DEVELOPERS

THE MASSES OF CITY DWELLERS have not yet followed the marine architect in his daring leap offshore. But they are going to. They are on the move, and the direction is oceanward.

The twofold Great Migration of our time is from the country to the city and from the interior to the coast. The common factor in both movements has been the Industrial Revolution. The promise of economic opportunity, prestige, variety, and amusement has been emptying the countryside and swelling the towns with job and fortune seekers, while at the same time the industries which lured them to the towns keep moving toward the seashore. Being near a seaport is good economics, for in spite of the development of rail, truck, and air traffic, sea-lanes remain by far the most important avenue of transport; maritime traffic, as we have seen, keeps growing. The seaside offers other advantages too, like unlimited cooling water and a waste dump that was long presumed to be all-absorbing.

While traditionally land-based industries are migrating shoreward, traditionally sea-based industries are also growing in the coastal areas, and new ones are being added. In the late 1960s, 84 percent of all marine exploitation (including fishing and fish farming, shipping, dredging, drilling) took place in coastal locations, 14 percent offshore, and only 2 percent in deep water.

Technological advances are changing these proportions to some extent, but as the Industrial Revolution penetrates the deep ocean space, it also intensifies the uses of offshore areas. The expansion of national jurisdiction in ocean space, with the impending establishment of a 200-mile Exclusive Economic Zone, is both a consequence and a cause of this development, and constitutes a new challenge to coastal management. Coastal-zone planning, legislation, and regulation will have to cover an area reaching 200 miles out into the sea and at least as far into the hinterland—up to the watershed. And they will have to cope with a host of interlocking problems—local, regional, national, and international. Among their concerns will be oceanographic research, coastal fisheries, open-spaces management, urban and industrial planning, water resources, flood control, tourism and recreational activities, harbor management, navigation, channel dredging, coastal erosion, aquaculture, fish farming, pollution, sanitation, customs, and surveillance—including naval operations, aircraft, satellites, and their land-based support.

"What is remarkable and unexpected, perhaps, is the apparently steady and seemingly inexorable drift of population toward the maritime frontiers, not into the zone of 200 miles distant from the sea or its lines of entry into the continents, but even within a few miles of the ocean—particularly along the fluctuating shores of the seas themselves," Norton Ginsburg writes ("The Lure of Tidewater . . .").

And it is not only industry that moves oceanward. Those who take refuge or seek respite from industry—tourists, vacationers, weekenders, retired people, the leisure class become leisure mass—also invade shores and isles. The number of international tourists climbed spectacularly, from a world total of just above 25 million "arrivals" in 1950 to 168 million in 1970—multiplying more than sixfold.

People's movements on the shores are like the wave movements beyond, with clashes and backlashes and tides and ebbs. We could use the Mediterranean or the Caribbean as an example, but the pattern seems to be fairly general and to reproduce itself, moving from the more-developed to the less-developed regions and altering landscape and seascape as it goes. The industrial urbanizer and the tourist—these are the shore developers, and they are definitely actors in the drama of the oceans.

As the scenario unfolds, industry moves oceanward, draining the countryside, draining also the villages along the coast. There is more money, presumably, in industry than in farming or fishing.

Then the tourists flow into the vacancies left by the farmers and the fishermen, and the countryside is restructured to their needs. Roads are built, and airports, hotels, restaurants, beach establishments, and night clubs. Yacht marinas are staked out; there is a mushrooming of bungalows and condominiums, shopping centers and specialty outlets for high fashion, local handicrafts, and water-sport equipment. All are built to maximize immediate profits and exploit the "natural resource."

The marine-recreation industries have grown into a multi*billion*-dollar business, comprising everything from bathing-suit fashions to children's beach toys, water skis, surfboards, rubber floats, canoes, snorkeling and scuba-diving equipment, underwater wristwatches, subaqueous photography and hunting equipment, and yachts and speedboats—to which small pleasure submarines must now be added.

With the exception of boating, which is ancient, the water sports galvanizing so much of the new marine tourism are all either of recent origin or else are recent reincarnations of much older sports.

Surfing, for example, was a sport enjoyed by the Incas and the ancient Polynesians. When Captain Cook came to Hawaii in 1778, he found the sport fully developed, and he marveled at the olo, the ritually and elaborately built surfboard of the chiefs and kings, fourteen to eighteen feet long and weighing 150 pounds.

The surfboard may originally have been a primitive boat, a means of transportation. Many sports, in fact, are playful developments of obsolete modes of transportation—rowing, sailing, and skiing, for example. Others are stylized developments of forms of combat and war—like most ball games—or they are straight and simple challenges to nature, like mountain climbing. Or, of course, they may be any combination of these. Surfing, which went into decline in the nineteenth century and was resurrected in the 1910s in Hawaii and California, is both a playful development of an outmoded form of transportation and a huge challenge to nature. Surfing, as John Severson says in *Modern Surfing Around the World,* means "harnessing, if only for a few moments, the tremendous power of the ocean. In surfing you merge with the wave, becoming part of its awesome force. . . . In the big surf you can come as close to danger and death as you want. It's up to you. And each year surfers tackle waves from twenty to thirty feet with only one purpose: to make the surf or beat the break—to take the hair-raising plummet down the face of tons of moving water—and hear the thundering crash as the wave breaks only feet behind you—to feel the solidity of the board under your feet as you glide out of the wave, away from the break, knowing that you have escaped and attained victory."

What the surfer sees is as exhilarating as what he senses and hears, and surfers have developed a unique technique of photographing the hollow of the wave before it breaks, and of photographing the world as it looks from inside the "tunnel" and through the spray. Curiously, it looks like the world seen from the moon.

The major surfing areas of the world now are in Hawaii, California, Florida, Australia, New Zealand, France, Mexico, Peru, and South Africa. If the present growth continues, surfing may become one of the top sports, and may even be included in the Olympic Games.

Skin diving, which got under way on a large scale only after World War II, seems to cast a spell on its practitioners, much as surfing does. Underwater diving and swimming and the direct contact with the underwater world are inebriating. One thinks one is dreaming, one is flying, one is in another world. To flutter over a living coral reef, its blue-green algae, sponges, spiny sea urchins, and elegant sea lilies, its clams and oysters, and the brightly hued, strangely shaped tropical fish weaving in and out—once we have fallen under the spell, we are enthralled forever.

Like the surfer, the skin diver has developed unique photographic techniques. They also share the thrill of danger, of defying nature. But the skin diver is, in addition, an explorer and a hunter. He may also help in conservation work, perhaps by combating the starfish that devour shellfish beds and kill off coral reefs; or he may engage in search-and-rescue work, or hire himself out for underwater construction or maintenance work on oil rigs or ship hulls. All this is "applied diving."

Both surfing and skin diving require considerable skill and courage. Water-skiing, a sport that became popular in the 1920s, is far easier. Anyone who can ski on land can water-ski. The land-skier, remembering his mountains, which are more like the surfer's waves, may even expect it to be boring. But it is not. On the contrary, it is breathtakingly thrilling, especially at speeds one can rarely reach on crowded mountain slopes.

There are, of course, all sorts of refinements, like trick-skiing on little disks rather than skis, or without any skis at all; or jumping off a jumping board: distances of over 130 feet have been reached by water-skiers jumping from a five-foot ramp. This kind of jumping can also be developed into flying, with a huge kite attached to the skier's back—there is no limit but the sky.

A strange new water sport was proposed in 1973, and that was water-walking ("Ariadne," *New Scientist,* January 10). The Central American basilisk lizard, popularly called the Jesus lizard, can actually do it. Humans need a wide-area shoe which will impact on the water but not sink very deep. "The simple solution is to make the shoe red hot," the *New Scientist* writer suggests. "Then it would never touch the water at all, but would be instantly cushioned on a layer of steam. This would also provide useful upward steam-pressure, giving a high degree of power assistance to the water walker." The proposed "water boot" would have a wide-flange sole separated from the foot and heated by a small propane burner. "The 'water bootman' strides smartly across the liquid surface, generating a dramatic hiss at each footfall, with the combination of steam power and frictionless gas support giving him a top speed of many knots."

One may want to take this with a grain of salt, but there is no doubt that sportsmen, and even nonsporting tourists, are moving deeper and deeper into the ocean, and in growing numbers. Underwater observatories and restaurants and tourist-sub trips through underwater marine parks are no longer the domain of science fiction.

In the Mediterranean, the number of tourists has more than doubled during the last fifteen years. As tourism increases, the resident population begins to replenish itself. People flock in to work in the booming service sector. Islands, coastal villages, and townships that had steadily lost population before the tourists arrived are growing. Congested, denatured, and polluting, the tourist center begins to sprawl, creating a belt along the coast. Campsites alternate with huge hotel complexes and frugal establishments where urban workers can spend two weeks in the sun.

Then the sprawling tourist belt hits the expanding industrial conglomeration. Factories rise right on the waterfront, and soon there is noise and dust, vehicular traffic, effluent discharge—and plain ugliness.

Disgruntled, the oldest residents—who are usually also the most aristocratic and own the oldest and most beautiful villas—pack up and begin their search for a new, unspoiled marine environment.

There the cycle starts again. Villas rise on unspoiled beaches, restaurants, hotels, and other facilities follow. The select few are followed by masses of tourists. And the same less-developed nations which can still offer unspoiled beaches to international tourism often also offer "pollution havens" to industries driven out of their own "developed" countries by stricter environmental legislation that cuts into high profits. Thus the pattern is repeated.

In the Mediterranean there has been a marked shift from the developed northwestern shores (Italy and France) to the less-developed southern and eastern shores (Greece, Tunisia, Lebanon), while other waves of tourism are inundating growing sectors of eastern Europe (the Adriatic, the Black Sea), the Caribbean and Mexico, and Africa. In spite of these recent developments, however, western Europe still hosts 75 percent of the world's international tourism.

In the Mediterranean, the tourist population mixes rather freely with the indigenous population, but in many developing countries the tourist population is kept segregated, either because foreign investors insist on excluding the natives from their own beaches (as happens, for instance, in the Caribbean), or because the host government wants to protect its population against the corruptive influences of international tourism while accepting it as a "necessary evil," as in Tanzania. Still other developing nations, like Tonga, have decided that this evil is not necessary, and frankly discourage international tourism.

Whether, in terms of social and economic development, tourism is a blessing or a curse has not yet really been fully explored. There can be no doubt that it creates foreign exchange, offers employment opportunities, strengthens the service sector, and enhances growth. But growth and development are not necessarily synonymous. There can also be no doubt that it distorts development, represses sectors such as agriculture, conflicts with industrialization, exaggerates inequalities between classes and between regions, and drives up prices. In global monetary terms, statistics clearly indicate that "at best, tourism at present involves only a minimal net flow of foreign exchange from developed countries to developing countries outside Europe, and at worst a net flow of foreign exchange to the developed countries" (John M. Bryden, *Tourism and Development*).

What is certain is that the growth of mass tourism, converging and conflicting with the growth of industry, has transformed the human geography of many maritime coastlines and is making a tremendous impact on the marine environment.

> . . . Beelzebub's grand Arsenal,
> where you meet so much Tumult,
> Thunder, Fire and Smoak, sometimes,
> that Old Nick himself cannot know
> which way to turn himself.
>
> Quoted by Oliver Warner,
> *Great Sea Battles*

CHAPTER TEN

THE WARRIORS

Man's changing role as a warrior in the oceans reveals the changing role of the oceans themselves in history.

Throughout most of historic time this role was secondary. War and wealth were continent-based and continent-oriented; navies battled for the possession of lands and to assure access to lands. Only during the second half of the twentieth century have the oceans come to be valued as repositories of wealth and marine areas begun to be battled over for their own sake.

The correspondence in time between this development and the final emancipation of the former colonies is only partly accidental. The two developments are in fact interacting, each being cause and effect of the other. Naval powers, unable to depend on overseas territories for cheap raw materials, turn their technologies toward the sea and the seabed. This in turn contributes to the independence of the former colonies—an independence, however, which may take the form either of marginalization and impoverishment or of development and participation. The outcome largely depends on how nations agree to manage the oceans from now on.

Clearly, the fundamental shift in the oceans' role must have a profound impact on the very nature of naval strategy, which is being transformed by at least three factors.

The first is the ever-closer interdependence between scientific research and warfare. The "conquest" of the secrets of the deep is, in many way, a military operation, and the tools employed—submersibles, underwater habitats, tracking devices, monitoring and forecasting systems—are applicable both to peaceful development and to war. The marine warrior thus plays a new and inextricable role in the drama of the oceans.

The second factor is that the oceans now hide the "second-strike capability" of the superpowers. Exposed on land to reconnaissance by high-flying spy planes and satellites, the most sophisticated weapons systems take refuge in the opaqueness of the ocean deeps. This again interlinks oceans and warfare in a new way. The dangers to the oceans, and to the world, need no explanation.

The third factor is still more potential than actual, but must nevertheless be reckoned with. Throughout most of history, naval warfare had included piracy. The pirates were a type of marine warrior; their ships and weapons were the same as those of legitimate navies; and the line of demarcation between legitimacy and piracy was often none too clear. The Phoenicians were pirates, and so were the Vikings, and pirates were kings over archipelagoes and islands from the Mediterranean to the Indian Ocean, from the Caribbean to the Red Sea.

What brought piracy down in the nineteenth century was "Big Technology," the kind of communications systems and weaponry that only nations could afford and that the outlaw could neither fight nor elude. But this happened so recently that the lawbooks have not had time to catch up with the change. The law of the sea still abounds with references to pirates and the punishment of piracy on the high seas.

Perhaps these laws might as well stay on the books, for there is bound to be a resurgence of piracy. The same "Big Technology" that makes nations strong also makes them vulnerable. Sabotage, terrorism, hijacking and skyjacking, and the diversion of radioactive wastes have become facts or fears of daily life on land. As the military-industrial-scientific complex penetrates the oceans, so will sabotage, terrorism, blackmail, and seajacking, giving a new dimension to naval warfare. Legitimate navies, private police forces in the employ of industrial companies, and neopirates will be involved in a situation that cannot be described either as war or peace. A new law for the oceans is the only alternative.

At the beginning, there was no difference between a ship and a warship. A warship was simply a means of transporting warriors. The first sea battle probably took place when a coastal nation that was about to be invaded sent out its own soldiers in boats to halt the invasion. The boats accosted the enemy, the fighting men boarded his ships, and hand-to-hand fighting ensued, just as it would have on land. When successful, this form of warfare had the obvious advantage of sparing the home country the ravages of war.

It soon became clear that oar power, which greatly enhanced maneuverability, was far superior to wind power. This was the beginning of specialization and the origin of the galley ship. The next important step in the specialization of the warship was the ram, invented in the Mediterranean sometime after 1000 B.C., during the transition from the Bronze to the Iron Age. The ram, Lionel Casson states, "must have had as revolutionary an impact as, say, that of the naval gun twenty-five hundred years later. A warship was no longer merely a particularly fast transport to ferry troops or bring mariners into fighting proximity with those of enemy ships; it had become an entirely new kind of craft, one that was, in effect, a man-driven torpedo armed with a pointed cutwater for puncturing an enemy hull" (*Ships and Seamanship in the Ancient World*).

Both the galley and the ram were basically Western developments. The Chinese used wind power, which they managed far more effectively than did the ancients in the West. At an early date, furthermore, they developed fast, treadmill-operated, paddle-wheel-driven craft. The introduction of paddle wheels was connected with that of ship armor, for armor could not be used with sail or with oar.

While the Chinese also used ramming, boarding, and hand-to-hand fighting, their naval battles left more room between ships. Hundreds of years before Christ, they already relied on the effectiveness of their crossbowmen. Their battleships were highly specialized in other ways as early as the sixth century B.C. Joseph Needham reports an interview between Ho Lu, the King of Wu who reigned from 514 to 496 B.C., and Wu Tzu Hsü, with the latter giving the following explanations to the king: "The classes of ships are named Great Wing, Little Wing, Stomach-Striker, Castled Ship and Bridge Ship. Nowadays in training naval forces we use the tactics of land forces for the best effect. The Great-Wing ships correspond to the army's heavy chariot, Little-Wing ships to light chariots, Stomach-Strikers to battering rams, Castled Ships to mobile assault towers, and Bridge Ships to the light cavalry."

Maritime fire power has a very long history. Its first documented appearance is on the warships of ancient Rhodes. As early as the third century B.C., the Rhodians had developed a way of "slinging containers of blazing fire onto poles that projected from the bows" (Philip Cowburn, *The Warship in History*). By the end of the first millennium A.D., effective firepower had been developed both in the East and the West. Casson describes the formidable "Greek fire" of the Byzantine navy during its heyday in the tenth century. Ships were equipped with one or more siphons, somewhat like cannon, consisting of a long, bronze-lined wooden tube connected to an air pump. Greek fire, made of liquid bitumen, or naphtha, and pitch extracted from coniferous trees, was poured into the tube, ignited, and pumped out through the muzzle. Pots of Greek fire were also shot from catapults; they exploded on impact, like grenades. Such catapults could also throw half a ton of lead over a distance of 750 yards.

The Chinese perfected a similar device on their formidable "Turtle" ships. These carried an animal figurehead "with a tube through which dense toxic smoke could be emitted, the result of the activities of chemical technicians hidden in the bows" (Needham). By early in the twelfth century, trebuchets for throwing gunpowder bombs were standard equipment on all Chinese warships.

In the West, gunpowder was introduced in the fourteenth century. In the sixteenth it completely revolutionized naval warfare. Until then, the object of naval warfare had been "to ram or scuttle your enemy, or to manoeuvre him into shallow water with your galleys and throw him out of action at the mercy of the sea or to get alongside and disable his tack and tackle, cut his halliards and cordage with shearhooks mounted on long poles and then grapple and board at the waist for a hand-to-hand encounter" (G. J. Marcus, *A Naval History of England*). The big gun brought the end of such close-fighting.

The greater distance between ships and the impossibility of communicating between one ship and another produced a curious trend in the eighteenth century. Unable to rely on oral commands, admirals began to develop elaborate sets of written instructions in accordance with which forces had to be deployed and action unfolded. At first these were drawn up for every battle, then they became permanent—an ideal deployment that ritualized and stylized the naval battle and made of it a work of art, a minuet of death. This was very much in the spirit of the eighteenth century. Gifted commanders could bring variations, or "amendments," to the ideal plan, just as a gifted composer had leeway to operate within the constraints of harmony and form. But the pattern was set, and both sides accepted it. Opposing fleets would face each other in perfectly formed lines, sometimes five miles long. The ships were rated according to the number and size of cannon they carried. The flagships of the two opposing admirals—"first-raters"—were equipped with 100 guns. They occupied the center of their lines. Then, left and right, came the second-raters, with 80 guns. Third-raters carried 70 guns, fourth (the most common) 54, fifth 32, and sixth 18 guns. The lighter frigates were kept in the back. Each movement, each action was carefully prescribed.

There was time for the perfecting of this style, for basically there was very little change in war and weapons technology from the sixteenth century to the nineteenth. But then change came.

Just as composers, painters, and writers broke through their formalistic constraints in the nineteenth century, so did the admirals. For them the liberating factor was improvement of communication, which made the written instructions unnecessary. This change began during the last quarter of the eighteenth century, with the introduction of an effective system of signaling.

The first simple code of signals, employing five flags in varying positions, had been devised in England as early as 1647. Then more and more flags were added and new positions invented, until, a century later, a nonsystem of the most bizarre complexity had developed, its colorful display accompanying the minuet of death danced by those gilded and baroquely carved ships of the line.

Not until the late eighteenth century was a more rational system introduced. Then a system was adopted that, employing combinations of three or at most four flags, enabled commanders to communicate 9,999 different items of information.

Then came the Morse code, and radio, and this communications revolution was accompanied by spectacular improvements in other aspects of warfare—steam power instead of wind power; iron hull instead of wooden hull; armor developing at an equal pace with the increasing efficacy of the shell-firing gun. Thus the magnificent "ship of the line" gave way to the modern battleship.

Then came the submarine, the torpedo, and the mine. Ships could be so far distant from one another that they could not be seen with the naked eye, but cannon could find their targets over a distance of more than 35,000 yards (the parallel development of long-distance communication and long-distance destruction is typical of the history of technology). The aircraft carrier became first complement, then rival, then successor to the battleship. Now the missile-carrying submarine, capable of delivering nuclear warheads to targets thousands of miles away, makes the techniques of World War II seem as antiquated as those of ancient Greece appeared to us in 1945.

Even a cursory look at some of the battles that constituted turning points in naval history may help us to visualize the role of the warriors in the drama of the oceans.

The most famous of the classical battles, fought with galleys equipped with three rows of oars (triremes) and a bronzed beak to be used as a ram, was the Battle of Salamis. There, in 480 B.C., the Athenian statesman and naval commander Themistocles, though having fallen behind in the arms race with the Persians, inflicted a decisive defeat on the fleet of King Xerxes.

To deploy his small forces most efficiently—he had only 271 ships—and at the same time prevent the Persians from bringing their huge fleet of more than 1,300 ships into play, Themistocles chose the narrows at Salamis. Hemmed in by rocks on either side, the Persians could be attacked in small groups. As wrecks began to accumulate in the straits, the Persians, trapped and confused, brought about their own destruction.

One of the mariners at Salamis was Aeschylus. In his drama *The Persians,* he conveys what must have been a tremendous personal experience through the words of a Persian messenger bringing the news of the calamitous defeat to the mother of Xerxes:

> The first rammer was a Greek
> Which sheared the great Sidonian's crest;
> Then close, one on another, charged the rest.
> At first the long-drawn Persian line was strong
> And held; but in those narrows such a throng
> Was crowded, ship to ship could bring no aid.
> Nay, with their own bronze-fanged beaks they made
> Destruction; a whole length of oars one beak
> Would shatter; and with purposed art the Greek
> Ringed us outside, and pressed and struck; and we—
> Our oarless hulls went over, til the sea
> Could scarce be seen, with wrecks and corpses spread.

Agrippa, Octavian's chief naval commander, who established Rome's supremacy in the Mediterranean, must have learned his lesson from Salamis. In the Battle of Actium (31 B.C.), where he won a crucial victory against the fleet of his opponent, Antony, he managed to keep his ships out of the narrow Ionian straits and in the open sea. His secret weapons were speed, maneuverability, and firepower. With his small ships, he outmaneuvered the rams of his clumsier and slower opponent, on whom he showered arrows around which oiled and blazing tow was wound. He also loaded rafts with combustibles, which he set on fire and turned loose in the direction of the enemy.

The last battle of triremes was fought more than 2,000 years after the Battle of Salamis, at Lepanto, where a 273-ship Turkish fleet confronted the 300 ships of the Holy League of Spain, Venice, and the Papal States. The Christian commander, Don John of Austria, half-brother of Philip II of Spain, gave detailed instructions before the battle started. Among other things, no one was to fire "until near enough to be splashed with the blood of the enemy."

The fight was much like a land battle, and raged all along the line. It lasted from the dawn of October 7, 1571, until nearly midnight. The flagships of Don John and Ali Pasha, the Turkish commander, met at the center. The Christian archers and musketeers opened fire. Three hundred rowers left their oars and, armed with pike and sword, attempted to board the Turkish ship, where they were swept back time and again. So were Ali's mariners when they, in turn, attempted to board Don John's ship. When they ran out of ammunition, they bombarded one another with oranges and lemons. The stalemate was broken at last by a Spaniard who rammed and boarded the Turkish flagship. All 400 Turkish fighting men fell in the hand-to-hand fighting that ensued, and Ali Pasha cut his own throat. Forty Turkish ships were sunk or burned, and 170 captured; some 15,000 Turks were slain or captured. The battle marked the end of Turkey as a Mediterranean sea power—and the end of the age of the galley. From this time on, the distance between fighting ships began to increase.

The British had developed a vast array of cannon, from heavy ones, which fired at short range, to the light culverin, which had a far wider range.

Philip II of Spain was fully informed of England's revolution in naval warfare, both by his spies and by a series of unfortunate encounters between his fleet and that of Francis Drake. Embarking on a frantic arms race, Philip built a fleet that was far more imposing than that of Queen Elizabeth, with ships that carried three times as many types of cannon as the English. But they were heavy and of short range, for Philip still believed in boarding and close-fighting. His instructions to his commander-in-chief, the Duke of Medina-Sidonia, read: "The enemy's object will be to fight at long distance in consequence of his advantage in artillery. . . . The aim of our men . . . must be to bring him to close quarters and grapple with him."

The Battle of the Armada, fought chiefly in the English Channel, began on the night of July 28, 1588, when the English sent eight fireships against the closely anchored Spanish fleet. This caused havoc among the Spanish, who, in haste to break anchor, confused their own lines and broke their formation. Many of their ships drifted onto sandbanks and rocky beaches. The English attacked at dawn, harrying the battered Spaniards with their long-range cannon.

On July 30 the wind changed, allowing what remained of the Spanish fleet to escape to the north, hoping to sail home by way of Scotland and the west coast of Ireland. The British, running out of ammunition, desisted from further pursuit, but wind and weather were also against the Spanish and some twenty-five vessels were dashed against the coast on the homeward voyage. In all, the Spanish lost 64 ships and 21,000 men.

"The trouble and miseries we have suffered cannot be described to your Majesty," Medina-Sidonia reported to his king. "They have been greater

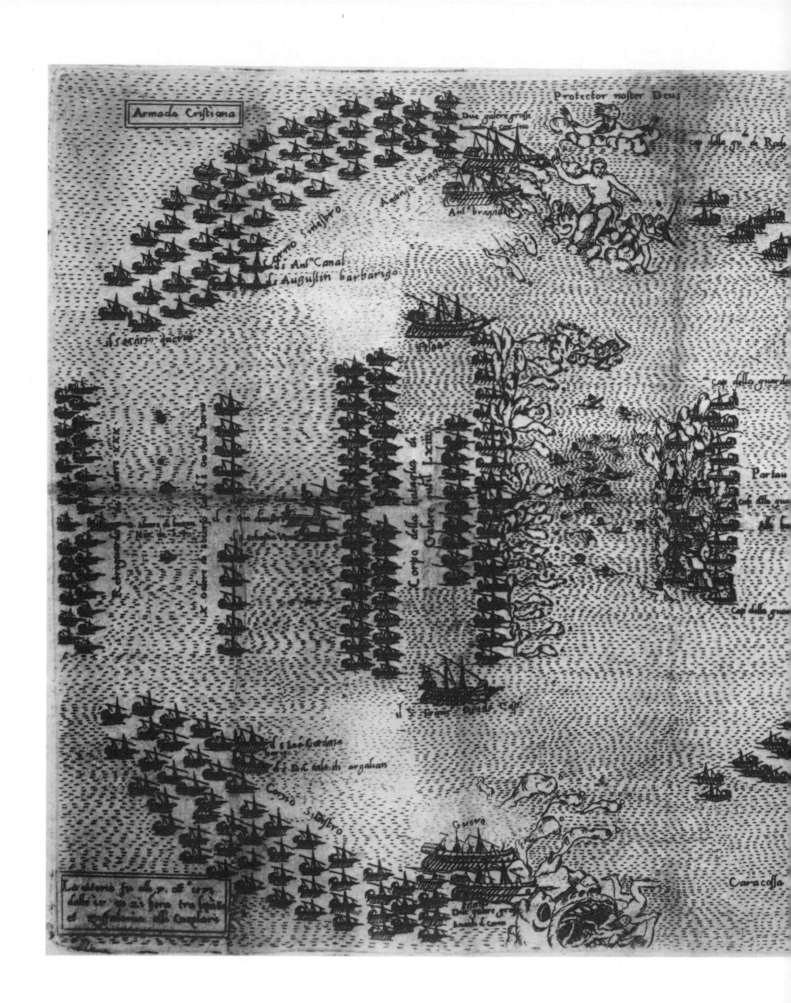

The Battle of Lepanto. A contemporary engraving showing the disposition of Turkish and Christian fleets, the number of ships involved, and the names of the commanders.
Museo Storico Navale, Venice

than have ever been seen in any voyage before." When Philip received the news, he said: "I sent my ships to fight the English, and not the wind and the waves. Praise be to God."

"That the Armada achieved nothing was ultimately due to the weather," Douglas Phillips-Birt comments in *A History of Seamanship*. "This must not disguise the fact that those July and August days of 1588 saw the new and the old ideas of applying seamanship to war in conflict. The old lost, and the lee shore of the Hebridean and Irish coasts, where the Armada's bones were cast, became their graveyard."

At Trafalgar (1805), Admiral Horatio Nelson did to the classical "instructions" what Beethoven did to the forms hallowed by Haydn and Mozart. He daringly broke all the rules, including his own. His plans for the battle that was to bring an end to Napoleon's designs for the conquest of England were precise but revolutionary. "No day can be long enough to arrange a couple of Fleets and fight a decisive battle, according to the old system," he said to one of his captains. "When *we* meet them . . . for meet them we shall, I'll tell you how I shall fight them."

> I shall form the Fleet into three Divisions in three lines. One Division shall be composed of twelve or fourteen of the fastest two-decked ships, while I shall always keep to windward, or in a situation of advantage; and I shall put them under an officer who, I am sure, will employ them in the manner I wish, if possible. I consider it will always be in my power to throw them into Battle in any part I may choose, but if circumstances prevent their being carried against the Enemy where I desire, I shall feel certain he will employ them effectively, and perhaps in a more advantageous manner than if he could have followed my orders. With the remaining part of the Fleet formed in two lines, I shall go at them at once, if I can, about one-third of their line from the leading Ship.

"This," Oliver Warner comments in *Great Sea Battles*, "was so daring, tactically, so unlike anything that had gone on before, and which would be so inappropriate as a model for lesser leaders, that it could be regarded as a plan to succeed only once, under particular circumstances, and with a fleet incontestibly superior not in numbers but in fortitude, gunnery, seamanship and training."

Napoleon's fleet, led by Admiral Pierre de Villeneuve and consisting of thirty-three ships of the line, was in fact numerically superior to Nelson's, which numbered only twenty-seven, including the magnificent flagship *Victory*. Each "first-rater" carried more than 100 cannon on its five decks. More than 900 men were needed to work the rigging and the guns of these ships.

Villeneuve was quite aware of the revolutionary tactics of his opponent, who, he wrote, "will not trouble to form a line parallel to our own and fight it out with the gun. . . . He will try to double our rear, cut through the line, and bring against the ships thus isolated groups of his own to surround and capture them."

When the two fleets were in sight of each other, Nelson assembled his captains on his flagship to give final orders, the gist of them being that the rear ships should do what they could as well and as quickly as possible, without regard to his own "order of sailing." They "scrambled into battle," one of the captains recalled later.

Nelson himself did what he had planned to do. He sent his fleet into attack in two columns, one led by Admiral Collingwood attacking the rear, the other by himself attacking the center. The battle did not last much more than four hours. "Partial firing continued until 4:30," the log of *Victory* reports, "when a victory having been reported to the Right Honourable Lord Viscount Nelson KB and Commander-in-Chief, he then died of his wound."

Then the age of steam and iron dawned. The encounter between two ironclad warships of the new era took place in the New World: the Civil War battle between the Confederate *Merrimack* and the Union *Monitor* at Hampton Roads, Virginia, in 1862.

The *Merrimack*, originally a wooden frigate, was scuttled by the Union forces at Norfolk in 1861. The Confederates raised and rebuilt her, and clad her in iron. "Inequality of numbers," Stephen Malloy, Secretary of the Confederate Navy wrote, "may be compensated by invulnerability. Not only does economy, but naval success, dictate the wisdom of fighting with iron against wood."
The *Merrimack* was cut down to the waterline and her sides were sloped, so as to be least vulnerable to cannon shot. She was then covered by two layers of two-inch-thick iron and fitted out with nine large guns. Slow—all her engines could achieve was a speed of 5 knots—and clumsy, she was put to work and tested in battle. Her mission: to break the blockade of Hampton Roads.

The first day was a great success. Two of the Northern wooden ships, the *Congress* and the *Cumberland,* were at anchor when the *Merrimack* arrived, looking like "a very big barn, belching forth smoke." The *Cumberland* opened fire at a distance of one mile, but to no avail. The ironclad proceeded undisturbed and opened fire at 500 yards. Then she rammed the *Cumberland,* which quickly sank. Next she turned against the *Congress,* firing at close range until that ship was totally ablaze.

The new was triumphing over the old.

It was not until next day that the *Merrimack* met her match. The Union Navy, too, had an ironclad, which sped to the scene. The *Monitor* was said to look like a "cheese box on a raft." The "cheese box" contained another bold innovation, a revolving turret carrying two guns. This presaged the end of the fixed broadside gun of the classical ship of the line, which had required so much more ship maneuvering and afforded so much less precision.

For seven hours the two ironclads were locked in close battle, occasionally with no more than forty yards between them. They battered each other like two fire-spewing dragons invulnerable in their thick skins. No life was lost. What was irretrievably lost was the age of the wooden sailing ship and its broadside guns.

The history of twentieth-century naval warfare begins with the Battle of Tsushima, fought in the Korean Strait near the small Japanese Tsushima Islands. Here, in 1905, the mighty Japanese Navy under Admiral Heihachiro Togo annihilated the Russian Baltic Fleet commanded by Admiral Z. P. Rozhdestvenski.

It was the first time that the recently invented wireless telegraph was used for battle communication. Togo's anchored ship was alerted by a wire dispatch when his scouts spotted the enemy fleet. And, as the range of communication had been extended, so had the range of guns. The Japanese were able to open fire at a distance of 6,500 yards with their twelve-inch guns, and it was these that were responsible for the Russian disaster. The Russian fleet was destroyed, and Russia disappeared as a naval power, not to reemerge until after World War II. Japan became the first Asian power since the demise of Chinese naval power in the sixteenth century able to "man and train a fleet equal to anything afloat," as Oliver Warner has described it.

During the following decade, the firing range increased from 6,500 to 19,000 yards, or eleven miles. This was the state of affairs at the time of the Battle of Jutland in 1916, when the German High Seas Fleet under Admiral Scheer and the British Grand Fleet under Admiral Jellicoe staged the greatest naval battle of World War I.

There were torpedoes and mines; a zeppelin airship flew overhead; submarines were in readiness; and there was even a British seaplane carrier, the *Engadine*. These new adjuncts of twentieth-century warfare, which were to transform the ocean from a two-dimensional to a three-dimensional battleground and modern sea battles into major holocausts, entered on the scene without much clamor. Their part in the big-gun-determined action was so tenuous that the zeppelin, flying high over the field of battle, was not even aware it was going on, and the *Engadine* was used only to tow away a disabled cruiser.

At the height of the battle, when Jellicoe's maneuvers were completed, "an immense line of battleships stretched to the horizon, by far the most powerful naval armament the world had yet seen," Warner writes, and the Germans perceived the arc stretching from North to East as a sea of fire.

Yet Jutland marks the beginning of the end of the great battle cruiser. There were no more big-gun naval battles in World War I. The German submarine took over, ravaging Allied shipping at a rate never seen before.

There were, of course, more great battleships in World War II, but they were outmaneuvered not only by the submarine but also by the aircraft carrier. At the Battle of Leyte Gulf in 1944, the carriers came fully into their own by defeating Japanese naval power and ensuring the American reconquest of the Philippines. "When the Battle of Leyte Gulf could be seen in some kind of perspective," Cowburn writes, "it was clear that the battle-ship had outlived her usefulness." The American battleships never succeeded in bringing their big guns to bear on the mammoth Japanese battleships *Yamato* and *Mushashi,* which met their end through divebombers and torpedo planes. After Leyte, battleships were used merely as escorts for carriers, or for bombardments in support of landings.

Many a nation has thought of itself as the hub of the world, and many a generation as the hub of history. Every generation develops its own eschatology in accordance with its own sense of guilt. Each generation perceives the break with the past more clearly than it does the continuity, and stresses the uniqueness of the moment more than multidimensionality of perspective. Yet uniqueness is defined by the future as much as by the past, and from a distance, whether in space or in time, gaps perceived by the close observer tend to disappear, and massive continuity, even monotony, takes over.

A close look at the warriors in the oceans of the late twentieth century nevertheless reveals pure horror. They have indeed made the oceans into Beelzebub's grand arsenal. Nations now can destroy the earth from the depths of the oceans. This is, in fact, the gist of the doctrine of the balance of terror and the availability of a "second-strike potential." If a nation feels so threatened that it wants to wipe another off the map with its atomic overkill, it will be deterred by the knowledge that countervailing power is hidden somewhere in the vastness of the oceans, and that intercontinental missiles with a range of many thousands of miles will be released by pushbutton. Thus the preemptive strike loses its appeal. This, at least, is the theory. What happens to the sense of security of the landlocked nations, without any ocean in which to hide their own "second-strike potential," or of nonatomic powers, does not enter into this "game," which is presently confined to two main players, the United States and the Soviet Union.

The United States possesses forty-one Polaris submarines, thirty-one of which are currently being converted to Poseidons equipped with MIRVed missiles. MIRV, the multiple independently targetable reentry vehicle, is one of the most important military inventions of recent years. One single ballistic missile fitted with a MIRV warhead can deliver, very accurately, one or more nuclear weapons to each of several different targets. This is achieved by a so-called Post Boost Control System or "bus," a guided missile traveling a set course including a number of targets, and releasing a "reentry vehicle" on each of its bus stops.

Each Poseidon has sixteen missiles with ten independently targeted warheads each. The United States thus has about 5,400 nuclear warheads currently deployed on board ballistic-missile submarines.

There is a great deal of public emphasis on the fact that, at our present state of technological development, these submarines are completely invulnerable. They defy any means of detection because the ocean is just too big and too dark. According to the most recent U.S. pronouncements their invulnerability is assured "at least until the late 1970s," but what the future holds, no one can tell.

Developments like the Lockheed S-3A carrier-based jet aircraft certainly dent, if they do not breach, the secrecy of the oceans. One such plane, the U.S. Navy claims, is capable of "sanitizing" 9,000 square miles of ocean on one single mission.

These submarine hunters, called Vikings, carry highly sophisticated computer-controlled detection and tracking systems, and are transforming the Navy's carriers into multiple warships, combining antisubmarine warfare with traditional attack missions.

Each Viking carries sixty acoustic listening devices that can be dropped at sea within ten seconds to pick up and transmit back to the plane all underwater sounds. The computer, making use of its memory bank, can sort out and identify information received from the microphones of thirty-one of these floating devices simultaneously.

The Viking's principal weapon is the Mark 46 homing torpedo, which, the Navy claims, is capable of overtaking the most elusive submarine known.

With such developments in antisubmarine warfare in existence or in the offing, work has already begun on replacing the Polaris/Poseidon by the so-called Undersea Longrange Missile System (ULMS), placing intercontinental ballistic missiles on new, large, quiet, deep-diving submarines with a deployment area at least ten times as large as that of the Polaris. About thirty ULMS are to be built, at a cost of about one billion dollars each (a Polaris cost about $300 million), and

186

they are expected to be effective until at least the year 2000.

The Soviet Union, according to *Statesman's Yearbook* (1974–75), has 120 nuclear-powered submarines. Of these forty-two are of the Polaris capacity and range. The Soviet Union is also building a carrier fleet. The first of these new ships, the *Kiev,* has been completed.

The United Kingdom and France each have a small number of ballistic-missile submarines.

Antisubmarine warfare relies on a great variety of weapons systems and sensors installed on the seabed or carried by submarines, buoys, surface ships, planes, and satellites.

The United States has led in the development of acoustic submarine-detection systems (sonar), installed on the seabed along both its East and West coasts and in various barrier zones guarding the exit of Soviet submarines into the deep ocean areas. Acoustic detection systems have been vastly improved during the last few years. As Sven Hirdman described it in "Prospects for Arms Control in the Ocean," these rely on "the pumping of megawatts of acoustic energy into the ocean so as to make the whole ocean basin ring like a bell and thereby betray the presence of submarines. For this, giant transducers are required which now are ship-borne but in the near future may be installed on continental shelves and in other shallow areas." Receiving networks, consisting of complex electronic equipment, are installed on the seabed, particularly on seamounts. All ocean space, including the seabed, is covered by these systems. The largest one ever conceived, the so-called Suspended Array System, is now under construction in the United States. It will cost about a billion dollars and consists of very large underwater structures, acoustic transducers, and electronic systems.

In many cases the seabed, submarine, surface, airborne, and satellite-borne detection systems deployed to monitor submarine and other military operations can be used—and are sometimes identical with those used—for environmental and other scientific observation and monitoring. Thus there are scientific by-products of military monitoring, and, inevitably, military by-products of scientific monitoring, in the ocean that "rings like a bell."

As for the weapons to be used against submarines after they have been detected, these include torpedoes, mines, missiles, and depth charges, with conventional or nuclear warheads. A particularly insidious torpedo mine is the CAPTOR, which can be dropped from a plane, a ship, or a submarine. It stays there awaiting the appearance of an enemy submarine, whereupon it explodes.

Another new development which may affect submarine and antisubmarine warfare is the construction of large air-cushion vehicles, the so-called Captured Air

Bubble (CAB) or Surface Effect Vehicle (SEV). These can achieve speeds of 50 to 100 knots, which means that they are faster than both submarines and torpedoes. The United States Navy has plans to build CABs of several thousand tons, which may be used as aircraft carriers in antisubmarine warfare. A type of large stable platform, constructed on the principles outlined in Chapter 8, may be used as a floating airfield for purposes of antisubmarine warfare.

The military are far ahead of anybody else in the creation of underwater stations. They have built such stations on seamounts at a depth of 6,000 feet, beyond the limits of the continental shelf. But that is not the end. Plans have been developed for such installations at a depth of 12,000 and even 20,000 feet. This makes most of the ocean floor available for military habitats. These stations may include nuclear reactors for their power supply. But not even the seabed is the limit. Habitats may be built into what is called the subsoil of the ocean floor—the term is inexact because there is no soil on the ocean floor, hence no subsoil. These are called Rocksites, in Navy language. They may be dug into underwater ridges or seamounts and serve for deep-ocean surveillance and submarine support. There is no law prohibiting their installation in the no-man's-land of the deep ocean.

The oceans were the scene of the creation of life. From the oceans life emerged to conquer the earth. And life in all its myriad forms still depends on the ocean for the larger part of its oxygen supply, and thus its very existence.

Now man is returning to the oceans in many guises, in many roles. There is a certain mythic quality in the fact that it is in the oceans that he has installed Beelzebub's grand arsenal, with which to destroy the earth.

"Ocean Venice," a marine city designed for a self-sufficient community
of 30,000 inhabitants. Courtesy Pilkington Brothers, St. Helens, Lancashire, England

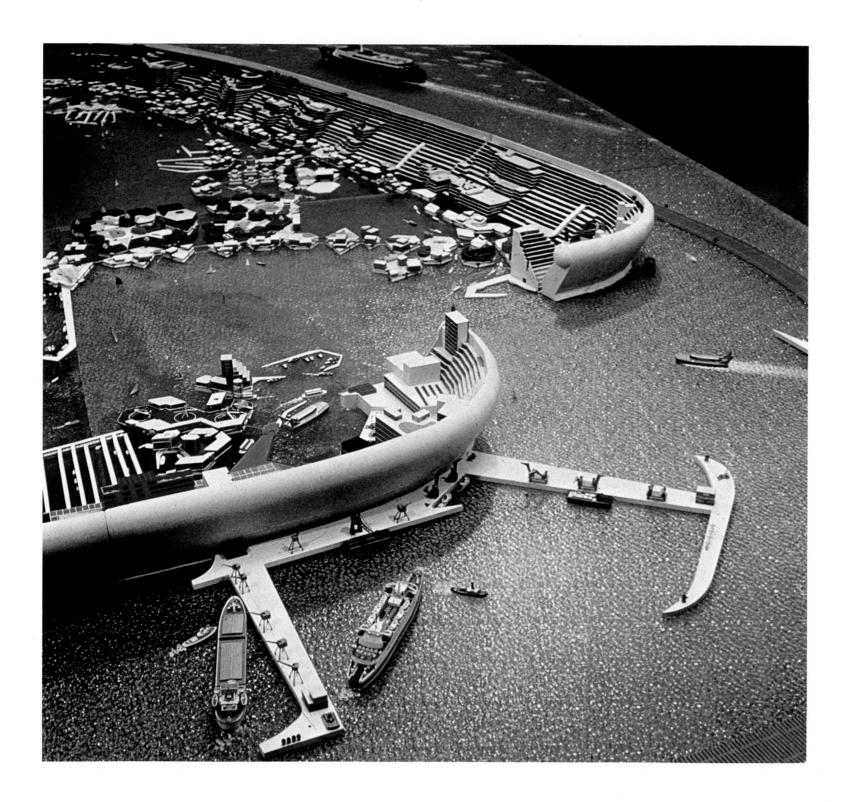

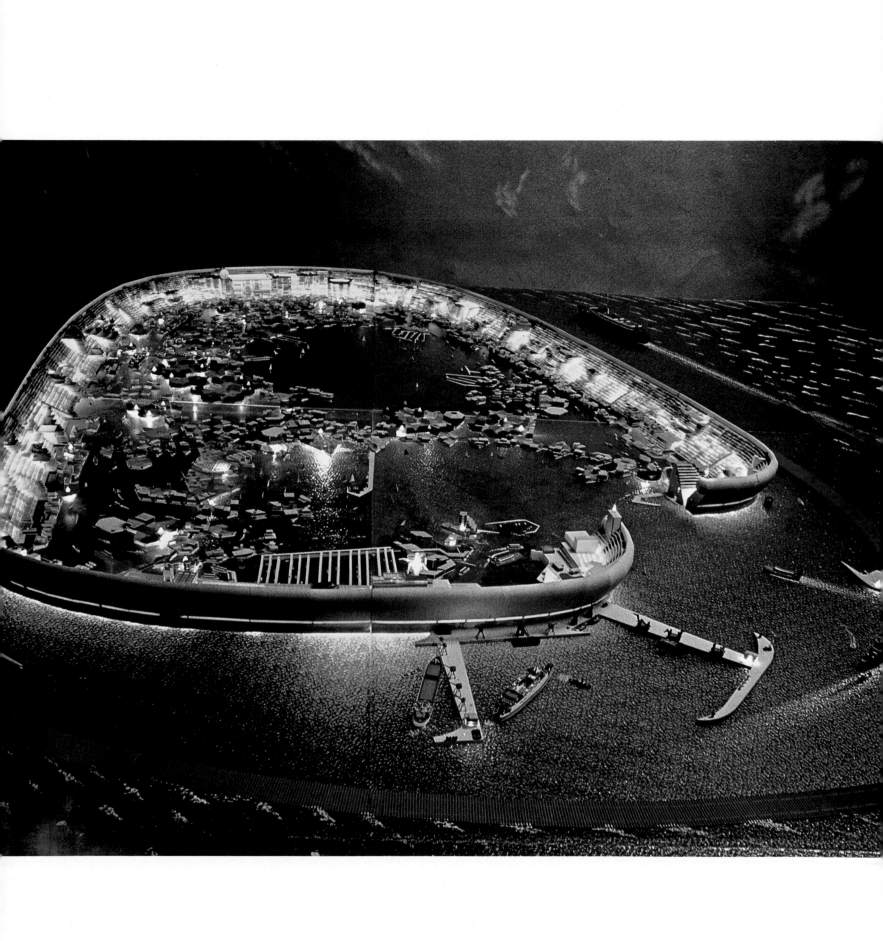

◀ John Craven. Hawaii Floating City Project,
a total systems design conceived to meet urban needs

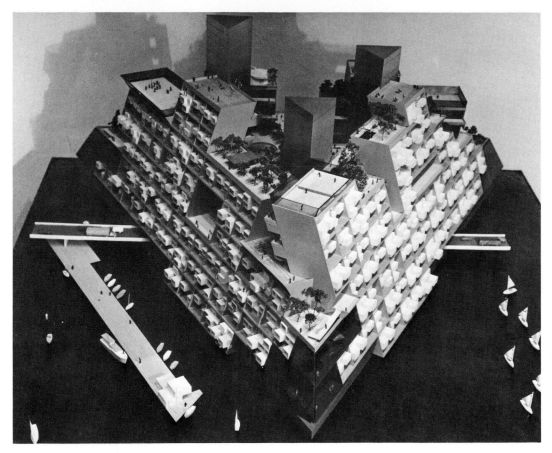

Buckminster Fuller.
This tetrahedronal island is a floating city

Plan for a multipurpose man-made island, part of the Sea Island Project.
Courtesy Bos Kalis Westminster Group, Sliedrecht, Holland

Surfing in Hawaii

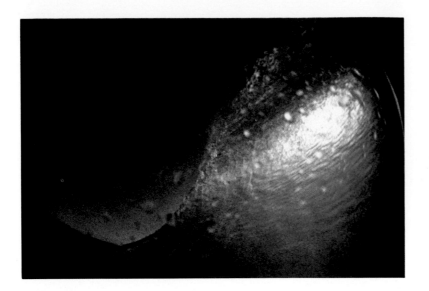 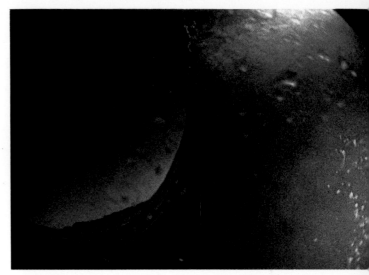

Frames from *Echoes,* a film made by George Greenough. With a
special f 1.5 lens that simulates human vision, Greenough was
able to capture the feeling of riding a wave while being
enclosed in the "tunnel" formed by its movement toward shore.
Greenough is seen entering the water with his surfing and
photographic equipment. Then, as the wave surges toward land,
he gradually becomes enclosed within it. The sun peers through
the back of the wave, the land races toward the surfer, and
finally the tunnel is fully developed as the top of the
wave rushes over his head in its downward plunge

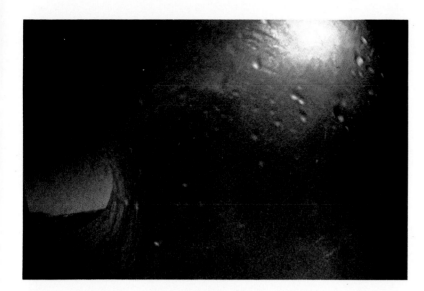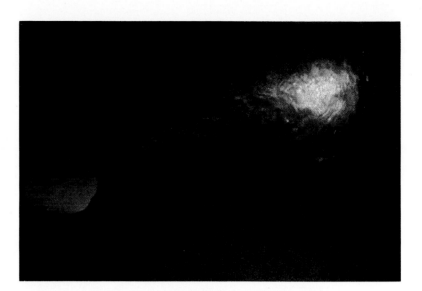

Venetian School. *The Battle of Lepanto.* Sixteenth century. Museo Correr, Venice

J. M. W. Turner. *The Battle of Trafalgar.* 1806. Oil on canvas, 68 x 94" Tate Gallery, London

Currier & Ives. *The Terrific Combat Between the Monitor and the Merrimac.* 1862. Lithograph. Museum of the City of New York. Harry T. Peter Collection

The Battle of Manila Bay. 1898. Lithograph

The torpedoing of the passenger ship *Falaba* by a German submarine. 1915. Lithograph

Above left:
The Polaris submarine
U.S.S. *Sam Rayburn*

Above right:
The launching of a Poseidon
missile from the submarine
U.S.S. *James Madison*

Right: The nuclear-powered
attack aircraft carrier
U.S.S. *Enterprise*

PART THREE

THE DRAMA

CHAPTER ELEVEN
THE DEATH OF THE OCEANS

WAR IS, OF COURSE, THE ULTIMATE POLLUTANT, and no one would be left to mourn the death of the oceans if even a small fraction of the infernal weaponry accumulating in Beelzebub's grand arsenal were to be exploded. But development for peaceful purposes may be equally destructive. Indeed, if present trends go unchecked, the oceans will surely die.

We have met the actors. There is no doubt that each of them has much to contribute. The trouble is that each considers himself the only protagonist. Each takes out of and puts into the oceans what profits him most today or at the latest tomorrow. Each refuses to see the long-range consequences, in space and in time—probably because they are so new. But the fact remains that we cannot behave with atom bombs in our hands the way we did when we were holding crossbows; the rules of traffic are different for vehicles speeding at 100 knots and for primitive craft straining to achieve three. Nor is the infinitely vast, unfathomed, bountiful and dangerous, mysterious ocean of a century ago the same as the mapped and charted, monitored, exploited ocean of today, contained by shores rocked by population explosions and technological revolutions. Either we adapt our behavior to the exigencies of this new ocean, or the results will be more and more devastating conflict.

Conflict, even in so-called peacetime, may take any of three forms: conflict between nations; conflict between nations and the international community; conflict between different uses of ocean space and resources.

Conflict between nations arises over territorial claims and the so-called limits of national jurisdiction. There are already numerous examples of this type: United States fishing boats have been seized by Peruvian and Ecuadorian patrols within 200 miles off their coasts—an area which these states claim as their territorial waters and the United States considers to be high seas. Iceland and Great Britain almost went to war—with battleships mobilized and fishing and coast guard craft being rammed and battered—over fishing limits. Iceland, depending for almost 90 percent of her livelihood on the sea around her, claimed waters up to fifty miles offshore, and Great Britain, needing room for her distant-water fishing fleet, insisted on holding to the traditional twelve miles.

These are the more clamorous cases, but there are many more. Between 1945 and 1972 there were 1,183 cases of conflict over fishery rights among major nations. An unpublished analysis by J. Sullivan and J. Zlonis shows that during this time period the Soviet Union, causing 208 cases of conflict, was numerically the main offender, followed by the United States with 168 cases. Japan, having actively caused only 85 conflicts, was the most frequent target: 258 times.

The People's Republic of China and the government of South Vietnam fought a brief war over possession of the Paracel Islands, for whoever possesses the Paracels possesses the oil on their continental shelf.

Libya and Malta, whose cooperation is essential for rational management in the central Mediterranean, almost broke off relations because of disagreements about offshore oil concessions. In the Aegean Sea, where oil-bearing salt domes have been discovered, tension is mounting between Greece and Turkey over jurisdiction, adding a materialistic base to the nationalistic drama of Cyprus.

According to recent estimates, there is more oil in the East China Sea and in the Yellow Sea than anywhere else in the world, including the Persian Gulf. But the "boundaries" are ill defined, and the claims of China, Taiwan, Korea, and Japan all overlap.

Conflict between nations and the international community may also be of a territorial nature. Such conflict is encouraged by the ambiguity of the international law of the sea with regard to the limits of national jurisdiction, and the United Nations has been working for many years to arrive at satisfactory compromises. But conflict of this type has also arisen over the uses, and misuses, of international ocean space and the violation of international treaties and obligations. The dumping by the United States Navy of nerve gas containers off the Bahamas created a conflict of this type. So did the French atomic explosions in the Pacific.

And this is just the beginning. If, abetted by loopholes and ambiguities in international law, nations extend their territorial claims with every new technological development, these conflicts will multiply and become ever more dangerous. The ocean floor of the North Sea has already been divided up among the coastal nations, and the same is bound to happen to the Mediterranean and other enclosed or semi-enclosed seas. It is an open secret that the big oil companies and their friends in or close to various governments are seriously discussing the parceling out of the big world ocean as well. Such a trend will almost surely create more problems than it will solve. Even supposing that no clear-cut old-fashioned war erupts over these concession wranglings: what about sabotage? What about weapons of terrorism in the hands of those who are left out of the division of the spoils?

Conflicts between nations and between nations and the international community are very obvious. Conflicts between various uses of ocean space and resources are far more subtle, pervasive, and difficult to solve within the present institutional frame-work. Yet their cumulative effect is tragic.

The fishermen are not as much of a threat to the other actors in the drama of the oceans as they are a threat to themselves. We have seen how, by overfishing, heedless international competition, and lack of positive interaction with other users, they are destroying their own resource.

But let us look at the others.

The damage wrought by navigation is not negligible. It is estimated that every year, about 7.1 percent of the world's fleet is involved in collisions at sea. Most of these occur in congested coastal areas, and many are caused by careless navigation or faulty nautical charts. James Dawson, a shipping insurance expert, says: "I doubt very much if you would be happy to take off from Dulles Airport for Europe in a Jumbo Jet, knowing that the navigator's map was made before the Wright brothers first became airborne. Ships all over the world's continental shelves are navigating with charts that in some cases have not been up-dated or even surveyed with lead and line, or single line echo sounders, since the days of steam, if not sail. The situation has worsened as each new vessel over 250,000 tons, drawing 20.5 metres (67′) when loaded, slides down the ways."

A number of accidents involve tankers and supertankers, such as the ill-fated *Torrey Canyon,* which discharged some 20 million gallons of crude oil into the sea; or the accident in San Francisco Bay, where two tankers collided and released more than 840,000 gallons of oil, contaminating large stretches of California beach; or the grounding of the *Metula* in the Strait of Magellan.

The probability of tankers and supertankers colliding or almost colliding with other ships is now such that one must assume a risk or near-risk situation approximately every four days, or nearly 100 times a year per tanker. That means about 10,000 risk situations a year, involving something on the order of 10^9 tons of oil!

Some of the latest tankers are atom-powered. The kind of risks such ships can create was demonstrated by the Japanese *Mutsu,* which developed a leak in her nuclear reactor, releasing radioactive waste into the sea. No port would accept the damaged ship. Shunned by other vessels, lacking conventional power, and with food running out, she drifted for nearly two months in the sea-lanes. The bizarre episode cost the Japanese nearly $40 million, including $4 million in compensation to local fishermen.

But ships pollute even when there are no accidents. Routine operations, such as tank cleaning and bilge pumping, release great amounts of oil into the oceans. It is estimated that over 5 million tons are deliberately discharged each year by ships of all types.

Oil production and consumption account for the rest of the pollution. Blowouts account for less than 10 percent of total oil waste in the ocean, but the results add up, and many accidents could be prevented if greater precautions were taken both with regard to equipment and training of crews for emergency action. This, however, is expensive.

Normal losses from offshore oil production amount to at least 100,000 metric tons a year, probably more. Runoffs from rivers, industrial wastes, and automobile wastes account for half a million tons. The total amount of oil discharged into the oceans in 1970 was about 2.1 million metric tons. Adding the possible fallout of airborne hydrocarbons on the sea surface, estimated at 1.8 million tons, the total amount of oil and oil products contaminating the oceans may be as much as 0.5 percent of the total world production. Production is estimated to double by 1980, reaching 4 billion tons per year. Thus the amount of oil entering the oceans may then be as much as 20 million tons.

This is indeed a nightmare for anyone who has been through an oil slick; sailed through a sea of oil, with the waves heaving their nauseating black crusts; swallowed an ocean breeze of petroleum, foulness, and death.

Today we know only some of the consequences of this scandalous waste. New consequences are constantly being reported as monitoring and research proceed.

Oil poisons marine life: filter feeders such as clams, oysters, scallops, and mussels die. Edible species of fish become inedible because of the presence in oil of the chemical benzo 3–4 pyrene, which is carcinogenic.

Oil disrupts the ecosystem, not only by the destruction of juvenile forms of marine life and of the food sources of higher species, but also in subtler ways which we are just beginning to understand. Communication and orientation among marine creatures often depend on a chemical "language"; messages related to mating, aggression, danger, and homing are coded in chemicals they emit. This chemical "language" makes use of hydrocarbons, among other things. Pollutants may interfere, either by masking these natural hydrocarbons or by mimicking the natural stimuli and leading to inappropriate responses. The homing of salmon, migration of tuna, or fish schooling or mating—all these may be disrupted, thus dooming entire species to extinction.

Oil may kill through contact poisoning, through coating leading to asphyxiation, or through exposure to dissolved or colloidal toxic components at some distance in space and time from the source. The intake of small, sublethal amounts of oil or oil products may reduce resistance to infections and other stress, and this may account for the death of many birds surviving the immediate exposure to oil. Other birds have died because oil, impregnating their feathers, expelled the layer of warm air which normally insulates their body temperature in cold water. Thus, when the bird enters the water, a large amount of body heat is lost, the metabolism is slowed down, and sudden death ensues.

The breaking down of oil by bacteria in the warmer oceans consumes large quantities of oxygen, thus affecting the reproduction of phytoplankton and algae. In enclosed seas like the Mediterranean, a film of oil even one molecule thick may possibly act "like a filter lens used on a camera, or in greater densities like an opaque curtain" that can "cut off and abort the photosynthetic process and hence the release of free oxygen" (Lord Ritchie-Calder, *The Pollution of the Mediterranean Sea*).

But it is not only with oil that the oilmen kill. They also kill with the explosive charges required by seismic surveying, which is part of all petroleum exploration. More than 10,000 wells have been sunk offshore during the last fifteen years, and although fish are killed only in the immediate vicinity of such surveying, the amount of loss adds up because, unfortunately, the fish are where the oil is—in the continental shelf area. Fishermen and oilmen are clearly antagonists in the drama of the oceans.

The energy engineers' contribution to the dismal scene is only at an initial and potential stage. Even though heat waste can be profitably applied to the culture of certain marine species, such as abalone, the unplanned heating of coastal waters that results from their use as coolants for nuclear and conventional power plants has caused harm in many places. Coral reef areas have been rendered unlivable for a number of species of fish. Many organisms in warm seas live so close to their upper limit of heat tolerance that any increase in temperature kills them off or drives them away. Increasing the temperature decreases oxygen solubility in the water and affects the metabolic activities of marine fauna and flora generally, and this may result in poisonous plankton blooms. We do not yet know how viruses and bacteria react to rises in water temperature, but it is quite possible that outbreaks of serious viral infections may result from environmental changes brought about by thermal pollution.

Even the most promising and nonpolluting energy installation in the ocean creates risks that must be considered. No one really knows the possible effects of large-scale thermal energy production on the heat load of the Gulf Stream, for example, and, given the enormous importance of the Gulf Stream for the climate of Europe, any tampering with it might give rise to international political problems.

Construction of huge dams to utilize the ocean's tidal power might seriously alter the tides by slowing down the rotation of the earth. This would make our days and our nights longer—we might find ourselves with a 25- or 26-hour solar day, giving unexpected reality to the plea: "Stop the world, I want to get off!"

Atomic power plants threaten the ocean environment in ways more dangerous than thermal pollution. We know that these plants produce radioactive wastes which can cause death, genetic damage, and cancer for many generations. These wastes are produced in direct proportion to the amount of energy generated. A power plant generating a million kilowatts of electricity annually produces radioactive by-products equivalent to those released by detonating a 23-megaton nuclear bomb. And this radioactivity persists for hundreds, sometimes even for tens of thousands of years.

Offshore atomic power plants are to be protected against natural disaster or ship collision by the largest breakwaters ever built. The breakwaters are designed, David Krieger points out in "Nuclear Energy and the Oceans" *(Pacem in Maribus III Proceedings),* to withstand "the greatest winds and waves experienced in the past century. This leads to the question: what if the winds and waves are greater this century than last?" What, we might add, if the plant is sabotaged, what if a hijacked plane is run into it?

If anything goes wrong with a floating nuclear power station, Henry Kendall of the Massachusetts Institute of Technology testified before a Senate committee, and "the remains of the reactor core and waste product melt their way through the reactor containment structure [something that will surely occur if the safety systems fail], contact between this material and the ocean water will cause the certain release of a very large quantity of solid radioactive wastes into the world's oceans. Such an event is a catastrophe of a kind the country has never experienced. There is in a large nuclear power plant, for example, enough strontium-90 to contaminate thousands of cubic miles of water. . . . Traces of this material would still be identifiable many hundreds of years after an accident. Strontium-90 . . . moves through the food chains and accumulates in the bones of human beings because of its chemical similarity to calcium. Because it lodges in the bones and undergoes radioactive decay there, it is linked very closely to leukemia." The thousands of cubic miles of contaminated ocean water, furthermore, recognize no boundaries. The disaster would be transnational.

Nor are the risks of disaster confined to sabotage or accidents during operation. Even if all goes well, there remains the problem of the routine disposal of the atomic garbage. Projects for shooting it into outer space or to the moon on rockets have been abandoned, in case of accidents during launching. Now NASA is seriously considering burying the wastes in the ocean floor. Canisters holding the deadly waste would be riveted deep into the basalt, below 20,000 feet of water and a thousand feet of mud, somewhere between Hawaii and Japan where the ocean floor is stable. This would have to become a regular operation, and the ocean floor would eventually be dotted with these canisters.

It may well be the safest place on earth, but that is not a very reassuring thought. Surely something will go wrong sometime, somewhere—while the radioactive waste is shipborne, or during its burial—and the consequences will be apocalyptic. Yet it is expected that the United States alone, which now has forty-four nuclear plants in operation, will have 1,000 in twenty-five years—and the rest of the world will be following suit.

Here we can see clearly that the peaceful development of this dangerous resource without adequate national and international safeguards is potentially as calamitous as its use in war. And the most likely primary victim would be the ocean and the living things therein.

The mineral miners are already doing their share of damage also. Dredging for sand and gravel stirs up sediments, and these clouds of suspended particles in the surface water may reduce the supply of sunlight to the phytoplankton and slow down photosynthesis. They may also interfere with the migration of fish. Materials dredged up in one place and dumped in another—for the clearing of a channel, for example— may settle on the seabed, smothering fish eggs, larvae, or even adult organisms, and destroying spawning grounds. Nodule dredging on the deep ocean floor may stir up and release hydrogen sulfide, a poisonous compound buried in marine sediments through natural processes. If nodules were not only recovered from the ocean but also processed in floating factories, as may well become economically convenient, great quantities of chemicals would be dumped into the sea.

Riverborne heavy-metal pollution of the oceans is quite considerable. A part of this is "natural"—minerals and metals are leached out of eroding mountains—but man outdoes nature on a grand scale. Thus the iron load that rivers carry into the oceans "naturally" is about 25 million metric tons per year. To this man adds another 320 million metric tons. The amount of lead carried "naturally" is 180,000 metric tons per year. To this man adds 2,330,000 tons plus another 500,000 tons carried through the atmosphere by wind and rain. (The United States alone burns

280,000 tons of lead a year in gasoline.) One would think mankind could make better use of these valuable metals than dumping them into the oceans, where they act as poisons.

The shore developers, urban developers, and industrial developers pour unbelievable quantities of pollutants into the oceans from New York Harbor, the Bay of Tokyo, the Bay of Hong Kong, Trieste Harbor, the mouth of the Po and the Rhine, and an infinity of other places. The disposal of sewage and of agricultural and industrial waste constitutes the gravest immediate danger to the oceans.

Can one imagine that *4 million tons of untreated sewage sludge* are dumped off New York Harbor each year, building up billions of tons and forming a dead sea of black, heaving, evil-smelling waste? The bacteria count multiplied by ten in the decade between 1960 and 1970. No animal or plant can survive in the filth; it rots the scales and kills the hapless fish that swims through it.

Like a nemesis, the sludge, thought to rest at a safe distance from human habitation, has somehow begun to move back toward shore. It has been sighted some eight miles from the Long Island coast. If it gets to the beaches, it could strike with hepatitis, cholera, and polio, ravaging the coast like a war.

Sewage pollution produces excessive growth of algae in coral reefs. The algae eventually smother the coral polyps, leading to a rapid decline in the animal community naturally associated with healthy reefs.

The combination of organic debris from sewage and industrial wastes can prevent the natural purification processes that normally kill pathogenic microbes contained in the sewage. Blooms (the overabundant growth of certain algae) become more frequent; nitrogen, phosphorus, and potassium are the main "nutrients." Of these, phosphorus has caused serious incidents. In Placentia Bay, Newfoundland, a release of phosphorus from a new phosphorus reduction factory led to massive fish kills, especially herring, and also endangered human lives.

When these discharges take place in enclosed or semienclosed seas like the Baltic or the Mediterranean, they may cause such a shortage of oxygen that finally only stinking marsh life can survive. When winds become impregnated with this foul seawater, they kill the coastal flora as well.

The most dangerous of the agricultural runoffs are the persistent pesticides or halogenated hydrocarbons, the best known of which is DDT, or chlorinated hydrocarbon. The world production of this chemical is about 200,000 tons per year. Because of its stability and mobility, at least half of this amount, and perhaps much more, enters the oceans, where it is taken up and concentrated by marine organisms. Of the 1.5 million tons of DDT produced since the end of World War II, two-thirds

A polluted river flowing through Tokyo and to the ocean

A Santa Barbara, California, beach soaking up oil from a recent spill

A common murre succumbs to one of the many forms of
death caused by contact with oil

A young Magellan penguin, victim of an oil spill
in Punta Tombo, Argentina

The brown pelican, a victim of DDT.

The thin shells of the brown pelican's eggs are the
direct result of DDT concentration through the food chain

California sea lion
carrying her premature pup.
Courtesy A. W. Smith,
from *Science*, vol. 181,
cover, September 21, 1973.
© 1975 by the American Association
for the Advancement of Science,
Washington, D.C.

Kumiko Matsunage, born in 1950,
became a Minamata disease victim in 1956.
Since then her existence has been a living death

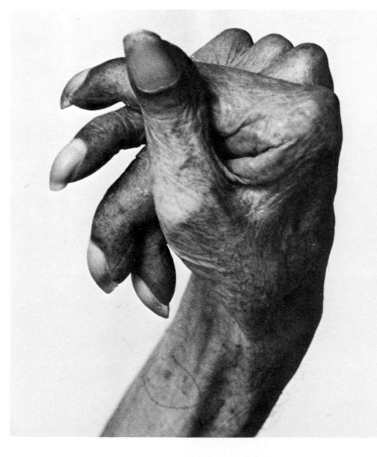

The deformed hand of Iwazo Funaba,
a victim of the dread Minamata disease.
He died in Minamata City after long suffering
as a result of mercury poisoning

may be adrift in the oceans. DDT is found in all oceans and all marine organisms—including Antarctic penguins, seals, whales—and in man as well.

Even in concentrations not sufficient to kill insect pests, DDT may inhibit the production of phytoplankton, ruin crustaceans and the eggs and larvae of some mollusks, interfere with ovary development and induce harmful changes in the tissues of mollusks and fish, and disrupt the schooling and feeding behavior of fish.

The pelican, a fascinating bird with a history stretching back some 30 million years, is dying out in many places. A group of biologists from the University of California at Berkeley recently examined a major pelican nesting site on the Santa Barbara Channel island of Anacapa. The results were shattering. As Joseph Jehl reports in "A Wonderful Bird *Was* the Pelican," the group examined 298 nests. "Nearly all were either empty or contained the remains of one or more broken eggs. Broken and discarded eggs littered the ground near most nests. Only twelve nests contained eggs; of these, nine held but one each. The nests were freshly lined with plant material; it was obvious that these small clutches were replacements for those that had been lost earlier in the season." When observers revisited the Anacapa colony two weeks later, all of the eggs were broken. All of the broken eggs were thin-shelled. Some were so thin that a slight pressure could distort them; others were hardly more than chalk-covered membranes and ruptured under their own weight.

It has now been well established that DDT is concentrated through the food chain in an amazing way: in the highest members of the food chain, such as fish-eating grebes or pelicans, the concentration may reach 80,000 times the concentration that was applied to a water area as a pest control. DDT causes eggshells to be thin and weak; recent studies have shown that the average pelican eggshell today is less than half as thick as those collected before World War II.

How this lingering oversupply of DDT affects human beings is not yet known, but recent studies on marine mammals raise some suspicions. In a report to *Science*, Robert DeLong of the National Marine Fisheries Service in Seattle, Washington, and his collaborators document how premature births in California sea lions are unmistakably linked to the accumulation of DDT and related chemicals in the blubber, liver, and brain tissue of apparently healthy mothers and dead or dying premature pups. The mating season of the sea lions in 1972 was between June 20 and July 20; the pups were due in mid-May of 1973. Miscarriages started in February. These early pups, born without fur, died soon after birth. The premature pups born in March, April, and early May were furred but lacked motor coordination; their breathing was short, and only part of their lungs had been aerated. They seemed to stay alive from a few hours to several days. Although the percentage of premature pups was difficult to establish, on San Miguel Island roughly 3.5 percent were premature; on San Nicolas Island it was nearer 10 percent.

One of the most pernicious industrial pollutants is mercury. Rivers have been carrying about 3,000 tons yearly into the oceans "naturally," but man has more than doubled this rate with his industrial wastes. Additional mercury, from the burning of oils and coal, reaches the oceans through the atmosphere. The total accumulating in the oceans may now be as much as 10,000 tons a year.

Mercury poisoning has caused untold suffering and many deaths. Minamata disease, which is contracted by eating mercury-contaminated fish, ravaged the Japanese working population in Minamata City, Japan, between 1956 and 1970. From there it spread to northern Japan, and its hideous effects have appeared in other regions of the world.

Patients afflicted by the disease display severe neurological symptoms, such as trembling, shouting like barking dogs, and a constant violent struggling. They suffer from concentric constriction of the visual field, loss of the ability to speak, disturbances in sensation and motor coordination, and convulsions. The dread disease has been transmitted to unborn babies by the passage of mercury through the mother's placenta, causing brain damage. Cats in the Minamata region who also ate the contaminated fish showed the same symptoms. Their affliction ended in a strange "cat suicide"—they jumped into the sea and drowned.

It took years before medical researchers, with very little help from the chemical factory that was responsible for the pollution, identified the cause of the disease. Demonstrations, even violent rioting, by thousands of fishermen took place before anything was done to remedy the situation.

Even in times of peace, the warriors contribute their share to polluting the oceans. This is seldom discussed in public, because convenient secrecy shrouds their activities. The United States Navy itself, however, felt the need to inquire into its own role as a polluter, and established the Navy Environmental Quality Program. "The program is directed toward the control and abatement of environmental pollution caused by the development, operation, and maintenance of Naval aircraft, ships and shore facilities. . . . Emphasis is given to developments that will reduce not only detrimental environmental effects but also improve military effectiveness" (*Pollution Control in the Marine Industries,* Thomas F. P. Sullivan, ed., 1972). The Navy, it was noted, pollutes with toilet sewage; its ships have traditionally lacked treatment systems on board. "In 1966 the Navy approached private industry to submit price proposals for the construction of such systems. From this effort and by coordinative efforts with other Federal agencies, the Navy currently has several shipboard sanitary waste treatment systems under observation. Conclusive results as to their effectiveness and suitability for shipboard use have not been determined to date." Nor have they yet.

Another area of ship pollution is solid waste—wood, paper, glass, rubber, plastic. Each man generates three pounds of waste per ship day. To these should be added boiler and incinerator emissions, oily wastes, and domestic wastes (galley, laundry).

Navy ships can be a source of significant oil pollution, Sullivan points out. The most obvious source is accidental spilling of fossil fuels such as jet-engine and diesel fuels. But what is even more serious, Navy ships have traditionally been designed to discharge oily wastes into the oceans by bilge pumping, fuel-tank stripping, deballasting of ballast water contaminated in fuel-oil cargo tanks, and fuel-tank cleaning.

In 1973, The Intergovernmental Maritime Consultative Organization (IMCO) adopted a Convention gently admonishing warships to deal with their sewage and fossil-fuel waste "in a manner consistent, so far as it is reasonable and practicable, with the present Convention." Obviously, the discretionary margin is conveniently wide.

Finally, there is the problem of the disposal of propellants, explosives, and pyrotechnics (called PEP materials), which contributes to the pollution of the oceans from the surface, by dumping; through rivers; and through the atmosphere, by open burning. The amounts involved are not always easily identifiable, but they are substantial.

Then, of course, there are also accidents. There may be collisions involving atomically powered navy units. An atomic bomb was once lost, as were two entire nuclear submarines, one American, the other Russian.

Thus the health of the oceans is attacked from the sea surface and the seabed, through rivers, and from the atmosphere. Organic and inorganic solids, liquids, and gases are accumulating much faster than the world ocean's water, flora, and fauna can assimilate them. Species after species disappears. Clearly, this trend must be halted. For while at present the wide-open ocean spaces are still relatively intact, and most of the damage has been done in coastal areas and in enclosed and semienclosed seas, somewhere along the line—and we do not really know where—there is a point of no return.

One need not be a romantic bemoaning the vanishing wilderness. One need not be a neo-Luddite, attempting to stop the machines. Yet one must recognize that what is being done to the oceans is irrational and unrealistic. While crying out over the presumed scarcity of resources, we dump them into the sea. With one hand we destroy what we build with the other.

It is not true that we must choose between economy and ecology, between man and nature. It is not true that we must slow our struggle against starvation. In the long run, a healthy economy requires a healthy ecology. Pollution is simply a measure of failures in our economic, social, and political systems. By destroying our environment, we are destroying ourselves. The list of endangered species includes Homo sapiens.

225

> . . . As I write, the sea whispers to me
> and I close my eyes. I am looking
> into a world unborn and formless,
> that needs to be ordered and shaped.
>
> Thomas Mann, *Tonio Kröger*

CHAPTER TWELVE

NEW LIFE FOR THE OCEANS

THE MARINE REVOLUTION CANNOT BE HALTED. Growing population pressure, food shortages, the near exhaustion of land-based resources, changes in the world's climate whose ocean-linked secrets we are barely beginning to fathom—all these factors conspire to push man into the sea, his last frontier.

Like all frontiers, however, the ocean frontier is closing, and closing rapidly. The seemingly infinite expanse is limited; the seemingly inexhaustible resources show signs of exhaustion.

The old concept of the freedom of the seas does not permit viable solutions to be found for the problems created by these developments. Unregulated freedom for intense, expansive, and conflicting activities is no more tolerable in the seas than it is on land. Yet while on land the nation-state has assumed the task of regulating and harmonizing conflicting uses of land and resources, the oceans are no-man's-land. So who is to regulate their uses?

The law of the sea, from antiquity to the end of the nineteenth century, was virtually synonymous with the law of navigation. Freedom of the seas meant freedom to navigate. This was set forth in the Code of Hammurabi almost 4,000 years ago and reformulated by other seafaring peoples since. Perhaps the most successful version was the Rhodian sea law, which remained valid for about sixteen centuries, from the second century B.C. to the fourteenth century A.D.

The Rhodian maritime code represents, in fact, a unique development. It dealt with joint ventures, charter parties, bills of lading. It regulated the behavior of passengers while on shipboard and established the liability of commanders and seamen in cases of injury to goods and persons. It fixed penalties for fraud, negligence, piracy, and careless collision. The most interesting aspect of the development of Rhodian sea law is that Rhodes—a small island in the Aegean Sea—was not a "superpower" in any sense. Its laws prevailed not because Rhodes was militarily powerful; they were voluntarily accepted by the seafaring world community because they worked.

Much of Rhodian sea law was later incorporated into Roman law and then into medieval law. The laws of Oléron, which set forth what was then common practice in the Atlantic Ocean, became the basis of much of English maritime law.

Freedom of the sea was asserted by disparate rulers at disparate times. Ivan the Terrible claimed freedom of navigation and called the oceans "God's road." Queen Elizabeth of England, answering Spanish complaints about Sir Francis Drake's depredations on the Spanish treasure fleet, is quoted as having said: "The use of the sea and the air is common to all. Neither can title to the ocean belong to any people or private person forasmuch as neither nature nor public use or custom permitted any possession thereof."

And the Hindus said: "This flood is the flood of life and belongs to all."

But long before—and long after—the famous dispute between John Selden and Hugo Grotius, the seventeenth-century advocates of the concepts, respectively, of an appropriable or closed sea *(mare clausum)* and the freedom of the sea *(mare liberum)*— there have been dissenting voices. Freedom of the sea, like other freedoms, has often turned out to be freedom for the stronger to ride roughshod over the rights of the weaker. Thus a spokesman for the proud Portuguese seafarers of the sixteenth century, João de Barros, interpreted this freedom in the following terms:

It is true that there does exist a common right to all to navigate the seas, and in Europe we acknowledge the rights which others hold against us, but this right does not extend beyond Europe, and therefore the Portuguese as lords of the sea by the strength of their fleets are justified in compelling all Moors and Gentiles to take safe-conducts under pain of confiscation and death. The Moors and Gentiles are outside of the law of Jesus Christ, which is the true law that all must keep under pain of damnation to eternal fire. If then the soul be so condemned, what right has the body to the privileges of our law? . . . It is true that the Gentiles are reasoning beings, and might if they lived be converted to the true faith, but inasmuch as they have not shown any desire as yet to accept this, we Christians have no duties towards them.

The Republics of Venice and Genoa subjected vast ocean stretches to their own sovereignty in the fourteenth and fifteenth centuries. And in 1493, just as the great age of seafaring and discovery was dawning, Pope Alexander VI, foreseeing a scramble for new lands and trade, issued the Bull of Demarcation dividing the Atlantic Ocean, between Portugal and Spain, along a "median line from pole to pole." To the west of this line Spain had exclusive rights of exploration and settlement; to the east of it, these rights belonged to Portugal. Ships sailing and trading under the flags of other nations had to take out licenses.

All this has a familiar ring today, when claims of extended national jurisdiction over the high seas and the ocean floor are being vigorously championed.

Throughout modern history, the limits of the territorial sea, which is the area over which a coastal state has sovereignty, were fixed at three miles from the coastline (this was originally based on the distance a cannon could shoot). Only during our own century did this limit begin to be extended to six and then to twelve miles—and then only for various special purposes such as customs control, hygienic regulation, or fishing. The first substantial change came in 1945, when, in the Truman Proclamation, the United States government claimed jurisdiction and control over "the natural resources of the subsoil and seabed of the continental shelf beneath the high seas but contiguous to the coasts of the United States as appertaining to the United States." The Proclamation continued: "In cases where the continental shelf extends to the shores of another State, or is shared with an adjacent State, the boundary shall be determined by the United States and the State concerned in accordance with equitable principles. The character of the high seas of the waters above the continental shelf and the right to their free and unimpeded navigation are in no way thus affected."

The Truman Proclamation became the basis of the 1958 Geneva International Convention on the Continental Shelf, which subjected the outer continental shelf to coastal-state jurisdiction out to a depth of 200 meters, or further if technology permitted the exploitation of its resources.

This so-called exploitability clause meant that the race to extend jurisdiction over the deep sea floor was on.

The assumption that the waters above the continental shelf are high seas and are in no way affected was a theory in need of a proof. For the water above the seabed is like the atmosphere above the land; and if the atmosphere above the land can be appropriated as "national air-space" and subjected to national sovereignty, why cannot the waters above the continental shelf?

In the 1950s, several Latin American nations announced that they were extending their sovereignty 200 miles out to sea. The notion of the "patrimonial sea," proposed by a conference of Latin American nations at Santo Domingo in 1972, and the similar concept of the "economic zone" proposed by the United Nations Conference on the Law of the Sea in 1975, are steps in the same direction, for both propose to extend coastal-state jurisdiction over resource exploration and exploitation, both on the ocean floor and in the waters above, to a distance of 200 miles from the shore.

But the oceans, and their problems, are too big to be dealt with by any one nation. If man is to live with the oceans, in the oceans, and from the oceans, he must adapt himself to the ocean system.

This system is one and indivisible. Geological structures extend, currents and waves move, species migrate across the high seas and the ocean floor. Everything in the oceans interacts with everything else, and the oceans themselves interact with the atmosphere and with the land.

To impose our rigid, conceptually land-based divisions on this flowing environment simply will not work, for static concepts cannot be applied within a dynamic system.

Division—even opposition—between man and nature, man and environment, is a static concept. Man and nature are a dynamic continuum, each a part of the other.

The social environment is clearly part of the total environment. Division—even opposition—between individual and community is a static concept. Individual and community are a dynamic continuum, each a part of the other.

Absolute ownership and "free enterprise" in the classical sense, postulating a rigid distinction between owners and nonowners, is a static concept. It is not applicable to a flowing, dynamic environment, where such a concept can be transformed instead into the right to use, in accordance with the requirements of all other uses, on the basis of the continuity between each and all, part and whole. All uses of the oceans must be integrated so that they enhance rather than destroy one another. New forms of interaction are needed among all users of ocean space and resources, among all actors in the drama of the oceans.

The full potential of ocean space and resources must be harnessed, not only to satisfy the needs of a still growing and developing world population, but because this is the only efficient way to minimize waste, and waste results in pollution.

Absolute sovereignty in the classical sense, postulating rigid boundaries between sovereign states, is a static concept. In a dynamic environment boundaries become fluid, and their geographic location matters less than does the functional content of jurisdiction—a number of decisions must be applied across boundaries, no matter where these are located, or they will not work on either side. New forms of cooperation are needed among local, national, regional, and global entities to cope with problems which are local, national, regional, and global. We must learn to do together what none of us can do alone.

New forms of coastal management are emerging on the local level that interweave the political, the scientific, and the industrial sectors, government and people, land and sea, in new ways. The time has come to interweave these with corresponding international forms of ocean management, for coastal management, vulnerable to environmental stresses beyond its control, cannot work without regional and global ocean management.

This does not require the establishment of a world state or superstate or the relinquishing of sovereignty. Such a notion is rooted in the static, land-based concept of division—even opposition—between the part and the whole. It presupposes a mechanistic concept of integration in which the whole gains at the expense of the part, or the part gains at the expense of the whole. The dynamic environment of the oceans imposes, instead, an organic concept of integration in which the part grows as the whole grows. Thus, there is no surrender of sovereignty. On the contrary, the growth of international community (the whole) fosters the growth of national sovereignty (the part).

Like everything else, the concept of sovereignty is transformed within the dynamic system and assumes a new dimension. This dimension is participation, participation by all groups in the making of decisions that directly affect them. Decisions made by one part of the system, whether by a nation or by an industrial or scientific enterprise, which affect the whole—such as the damming of an ocean basin, the opening of an isthmus, or the pollution of the ecosystem—destroy the sovereignty of all other groups that are forced to undergo passively the consequences of such decisions. To exercise their sovereignty, then, all nations must participate in the making of decisions that affect them. Since the problems we are faced with in the oceans are transnational and multidisciplinary, so must be our decision-making and planning.

On November 1, 1967, the Delegate of Malta, Ambassador Arvid Pardo, rose in the First Committee of the General Assembly of the United Nations to introduce an item on the Agenda: "Examination of the question of the reservation exclusively for peaceful purposes of the seabed and the ocean floor, and the subsoil thereof, under-lying the high sea beyond the limits of present national jurisdiction and the use of their resources in the interest of mankind." He drew the attention of the Assembly to the vast riches hidden on the deep floor of the world ocean which the technological revolution was rapidly making accessible to exploration and exploitation, and which did not belong to any nation. He pointed to the dangers of a military competition to dominate the deep seas. He saw a race developing to carve up the no-man's-land of the ocean floor in a manner that would give rise to acute conflict and pollution, and recalled the history of the African continent, which had been carved up by the colonial powers in a similar way.

Ambassador Pardo explained how the old law of the sea, based on the sovereignty of coastal states over a narrow belt of ocean along the coasts and on freedom of the seas beyond this, was being eroded, and suggested that a new concept, the common heritage of mankind, take its place. He stressed the ecological unity of ocean space and the interactions between all areas and all uses of ocean space. In conclusion, he suggested that the United Nations General Assembly declare the seabed and its resources beyond the present limits of national jurisdiction a common heritage of mankind, elaborate a set of principles to govern activities relating to the seabed, and then proceed to negotiate a treaty which would both clearly define the limits of the international seabed and create a new type of international organization to administer and manage its wealth for the benefit of all mankind. The common heritage of mankind would be used for peaceful purposes only, thus excluding the arms race from an area that comprises over three-fourths of the surface of the globe.

Few speeches heard at the United Nations have triggered as much activity as Arvid Pardo's address. A committee of thirty-five nations was formed to study the question and make recommendations to the General Assembly. A year later, this committee was enlarged and became the permanent Committee on the Peaceful Uses of the Seabed and Ocean Floor Beyond the Limits of National Jurisdiction. In 1969, a resolution prohibiting the exploitation of the international area prior to the establishment of the new regime and a resolution declaring the 1970s the First Decade of Ocean Exploration were adopted.

The United States and the Soviet Union submitted proposals for the demilitarization of the seabed, which eventually resulted in the 1971 Treaty on the Prohibition of the Emplacement of Nuclear Weapons and Other Weapons of Mass Destruction in the Seabed and Ocean Floor and the Subsoil Thereof.

In the autumn of 1970, the XXVth General Assembly of the United Nations adopted a Declaration of Principles governing the seabed beyond the limits of national jurisdiction, which elevated the principle of the seabed as a common heritage of mankind to a norm of international law.

The General Assembly also took another important step in 1970: it decided to convene in 1973 a conference on the law of the sea. The first substantive session was held in Caracas from June through August 1974; a second was held in Geneva from March to May 1975. The great issues before the Conference were summarized by

Luis Echeverria, President of Mexico, in Caracas in July 1974:

> Man's entire attitude with regard to the sea must change. The dramatic growth
> of the world's population, and the consequent increase in demand for food from
> the sea; the expanding industrialization on all continents; the congestion of
> populations in coastal areas; the intensification of navigation and the ever more
> frequent deployment of supertankers, containers of liquid gas, and nuclear-
> powered vessels; the increasing use of chemical substances which eventually end
> up in the seas—all these are factors which impose the necessity to regulate
> globally, to administer internationally, the uses of the oceans. Every day there
> will arise new and greater conflicts between different competitive uses of the
> oceans, conflicts which no nation will be able to resolve alone.
>
> There is furthermore a constant interaction between the multiple uses of
> the oceans. The exploitation of seabed resources may affect the utilization of the
> superjacent waters and vice versa; activities in international areas and in
> national coastal zones affect one another mutually; and the sea in its totality,
> and the atmosphere above it, form one ecological system. All these interactions
> demand a global and integrated vision and treatment of the marine environment.

No matter what the final outcome of the Conference, there can be no doubt that it is
the largest and one of the most important international conferences ever held. Should
it fail, the breakdown of international trade and scientific research would be
accelerated. The development of food, mineral, and energy resources would be
frustrated for a world population whose growing needs can no longer be satisfied by
land-based resources only. And the degradation of the marine environment would be
speeded. Conflict would be increased, and the arms race, given free rein in ocean
space, would add to the danger of total annihilation.

Let us look at the minimum requirements for rational development of ocean
space and resources with regard to each of our actors, in the order in which we
introduced them in the drama.

Most traditional commercial fisheries operate within 200 miles of a coast. These
will undoubtedly be conserved and developed by coastal management under national
jurisdiction in the so-called Economic Zone that is likely to be one of the results of
the Conference on the Law of the Sea. It is an illusion, however, that fisheries can
be managed successfully within national ocean space, no matter how large that space,
if the traditional "freedom of the high seas" reigns in international ocean space,
leaving nations and their industries free to overfish and to pollute beyond the limits
of national jurisdiction. Rational fisheries management in national ocean space
requires equally rational management for international ocean space, with the two
management systems cooperating and interacting.

Many fisheries could be developed on a regional level. Plans for regional development are under consideration for the Caribbean, the Mediterranean, and the Baltic.

Where highly migratory species and the resources on which they depend for their life support are involved, and with regard to highly mobile distant-water fishing fleets, whose range of operations transcends regional limits, universal criteria must be established and then constantly reevaluated on a worldwide basis. Trends must be projected and integrated. Some items may illustrate the range of issues that ought to be dealt with:

1. Conservation of overfished species by a rational reduction of local and distant-water fishing fleets (aided, during these next few years, by rising fuel prices);

2. Development of fish farming and aquaculture, especially in coastal zones;

3. Improvement in the harvesting of the abundant "lower trophic levels" of marine organisms, such as plankton and krill, and of technologies to process these into food for human consumption;

4. Development of the harvesting of midwater animals such as squid;

5. Development of equitable distribution, with special regard to the needs of undernourished, nonindustrialized populations;

6. Research into the interaction between fishing and other uses of ocean space and resources, for example, between fishing and oil exploration and exploitation; between fish trawling and cable laying; between fishing, fish farming, and the industrialization of shorelines; between commercial fishing and sport fishing;

7. Establishment of priorities, based on this research;

8. Establishment and monitoring of permissible levels of discharges into the sea; introduction of nutrients; transplantation of marine animal and plant species from one region to another; other changes in and manipulations of the ocean environment.

Wherever possible, the development of living resources should be integrated into multipurpose processes, so that energy can be generated, water desalinated, minerals produced, and waste recycled at the same time.

Although many fishing conventions, treaties, and commissions already exist, the problem is to integrate them into a sensible whole and to eliminate loopholes and contradictions. All these arrangements should be linked with the Committee on Fisheries of the Food and Agriculture Organization of the United Nations (FAO), a body which at present lacks an independent scientific and managerial capacity and should be restructured and strengthened to assume the regulatory, planning, and managerial functions required by technological developments. If the present confused and anarchical situation is permitted to continue, traditional fisheries will be depleted in the foreseeable future, new technological advances and the development

of unconventional fisheries and aquaculture will benefit only the rich and technologically developed nations, and waste, irrationality, and the degradation of the
marine environment will accelerate. If a rational regime for the development of the
living resources of the oceans is established, food from the oceans could be multiplied
manyfold and could make a vital contribution toward solving the staggering
problems of world hunger.

Just as for the fisherman, anarchy is no longer tolerable for the sailor. The traditional
concept of "freedom of navigation" cannot be applied to vessels whose speed and
size and cargo have multiplied so dramatically; nor to vessels which must operate in
industrialized oceans in which bottom and surface installations of all kinds are
proliferating; nor to vessels which, whether by accident or through routine operations,
are capable of polluting vast stretches of ocean space. "Freedom of navigation" for
such vessels is like applying the rules for horse-and-buggy traffic to rush-hour traffic
on the Place de la Concorde. It clearly will not do. Traffic lanes and other safety
measures are needed in congested areas and in straits; warning systems and rescue and
emergency services are needed for surface and underwater vehicles; standards must
be set for the prevention of pollution from ships and ship construction.

Such work is presently carried out by the Intergovernmental Maritime
Consultative Organization (IMCO), but IMCO's work might be integrated with
cartographic and lighthouse services, for example, and with the regulation of port
facility, superport, and artificial island construction as well as with the work of
the International Labor Organization (ILO) with regard to the training of crews.
IMCO is already in a process of transformation, widening both its membership and
its functions, but it still needs some restructuring before it can gain the full
confidence and cooperation of all nations, rich and poor, large and small. An
international licensing system for international navigation may have to be adopted,
to cope, among other things, with the inconvenience of the "flags of convenience," that
is, the registration of ships by states offering the most permissive safety standards and
financial conditions. Eventually, IMCO might acquire an operational arm, an International Sea Service which would sail vessels under the United Nations flag to deal
with rescue missions, environmental emergencies, and international law enforcement.

No organization now exists within the United Nations system to tend to the needs
created by the oilmen, energy engineers, and miners. The oilmen will operate, for
the foreseeable future at least, within the 200-mile limits of the Economic Zone
and thus be under national jurisdiction. Yet the effects of their drilling and spilling,
of the construction of huge underwater storage tanks, of the transportation of oil
through pipelines or tankers are transnational and must be regulated internationally.

The same applies to the establishment of safety standards for onshore, offshore, and floating atomic power systems and energy production from tidal, ocean current, and ocean thermal production plants. For all of these, if left uncontrolled, can hamper other uses of the ocean environment and even alter this environment in unpredictable and dangerous ways.

The miners operate for the most part under national jurisdiction, but the nodule miners harvest and process their manganese nodules from the deep floor of international ocean space. It is for them that the Conference on the Law of the Sea is establishing the International Seabed Authority, to manage the production and marketing of the minerals and metals contained in the nodules for the benefit of all nations. Safety standards for all miners must be developed. The impact of mining operations on the environment and on other uses of the oceans and their resources must be studied. World requirements must be projected, and the interaction between land- and ocean-based mining must be considered, with production plans taking into consideration the needs of nonindustrialized countries.

In many ways, the International Seabed Authority will be a new type of international organization. For the first time in history, natural resources located beyond the limits of national jurisdiction will become the common heritage of mankind and be managed by an international organization. This organization must be so structured that it can perform its managerial functions effectively and equitably for the benefit of all peoples. If it is successful, it may, in turn, provide some guidelines for the restructuring of the international organizations charged with managing various other uses of ocean space, such as the Committee on Fisheries, IMCO, and the Inter-Governmental Oceanographic Commission of UNESCO (IOC). IOC is the organization presently responsible for coordinating international oceano-graphic research. Like the other organizations, it lacks independent scientific capacity and the capacity for effectively transferring technologies from the developed to the developing countries. If IOC could be restructured and strengthened, and perhaps provided with an international oceanographic institute to advance the interna-tionalization of research, to train experts, especially from the poorer countries, and to transfer technologies, we would have four basic organizations dealing with the oceans: The Committee on Fisheries for the development of living resources; the International Seabed Authority for the development of nonliving resources; IMCO for navigation; and IOC for science and technology. The next step would be to link these four by some kind of integrative machinery so that their activities could be harmonized, activities not covered at present by any international organization could be regulated, the ocean regime could be linked with the United Nations system, and cooperation between national and international management systems could be articulated. Such a system would provide the global and integrated vision so sorely needed.

Only such a system could cope with the new and revolutionary actors we have seen making their entrance into the drama of the oceans.

The architects pose the problem of the regulation of artificial islands, whether floating or stationary, and of marine cities and parks. The location of such new territories must be approved, in accordance with safety standards and to avoid interference with established shipping lanes and other uses of ocean space and resources. If they are moving on the oceans, their routes must be determined. Their impact on the environment must be studied. The limits of their "territorial sea," their fishing rights, and their responsibilities for pollution controls, among other things, must be established. Certain marine parks and cities might come under the direct jurisdiction of the ocean institutions and be used for world community purposes, as international research centers or as bases for international production systems. The inhabitants of others might enjoy the privileges and bear the responsibilities of dual citizenship, participating in the political processes of their homelands as well as in the planning and management processes of the international ocean institutions.

The shore developers obviously operate under national jurisdiction, but many of their activities have transnational effects which require international regulation. These include atmospheric and river pollution, thermal pollution, harbor construction standards, functional zoning, international tourism and its harmonization with other uses of coastal zones. They might well also include the regulation of engineering projects such as dams, isthmuses, and canals, the effects of which, either on the environment or on populations, are transnational. Archaeological treasures from underwater sites may be of interest to the shore developer as tourist attractions, but their recovery may have international implications. As technology advances, such treasures may be recovered even from the deep international seabed. Wherever they are found, archaeological sites and wrecks must be protected from plunder and must be treated as part of the common heritage of mankind.

The final group of actors, the warriors, will be the strongest force opposing the new international order, for disarmament and arms control in the oceans depend on the political climate on land. Nations will not arm on land and disarm in the ocean; quite the contrary. Certain impulses can come from the oceans, however, and serve to advance disarmament and arms control. National navy and coast-guard units may be designated for service for international community purposes. Certain military activities or technologies may be excluded from particular ocean areas. Thus, for example, the installation of atomic weapons and other weapons of mass destruction has already been prohibited on the seabed. Such limitations might be extended to other areas. The Indian Ocean or the Mediterranean Sea might be declared zones of peace; the Latin American denuclearization treaty might be extended to cover the Caribbean. Extraneous nations might renounce the deployment of their naval forces in those areas, while coastal nations might reduce theirs. Tracking devices and monitoring systems on the seabed, the sea surface, on submarines, and on satellites might be internationalized and designed for multipurpose use (monitoring events and developments with an environmental impact as well as the movements of ships and submarines), which would reduce their cost and increase their efficiency. By and large, military activities will find themselves increasingly restrained as the oceans are used for more and more other purposes, whether national or international.

The new ocean "institutions" or "communities" that we have envisioned represent a new type of international organization, partly political, partly economic, partly scientific in nature, and bridging these three realms of human endeavor in a new way. They combine features of government and industry and are a union of workers and producers, a system of organizations neither national nor international in the traditional sense but articulating both spheres of action in a new way.

If in the remaining years of this decade we succeed in creating such a system of organizations to solve the problems of the oceans—and only such a system can solve them—we will in fact have done more than solve the problems of the oceans. We will have built a model for international organization in the twenty-first century. The oceans are our great laboratory. If nations can harmonize their activities in ocean space, they can cooperate in the areas of outer space and satellites, weather control and modification, energy production and distribution.

Already, other areas are bidding for attention. The icy wastes of Antarctica have been governed for many years by a Treaty—in some respects a precursor to the nascent Ocean Treaty—reserving the region for peaceful uses and scientific cooperation. Now important minerals—copper, nickel, and oil on a large scale— have been discovered, and the question has been raised: Whose is this wealth? Will it be subject to conflicting national claims, bringing to an end the beneficent rule of the Antarctic Treaty? Or will the Antarctic remain a zone of peace, to be developed in international cooperation and under international authority for the benefit of all? Clearly, the outcome of the Conference on the Law of the Sea is bound to affect the fate of Antarctica and to teach us lessons of great practical importance for the management of all resources and technologies of transnational impact.

All of this will undoubtedly take time—and more mismanagement, waste,

Air and sea combine to form a gently swirling cyclone

An elephant seal basks in the sun

California sea lion (*Zalophus californianus*)

A small boy greets the incoming tide

Sailing in San Diego Harbor California

Sunset on the ocean

The rise and ebb of the tides shapes the shoreline cliffs ▶

pollution, and human suffering. Things may well get worse before they get better. Or, even more likely, they will continue to do what they have always done—they will get worse and better at the same time. Yet in facing the great challenge of the oceans we have a chance to do better than in most other areas. Vested interests are as yet less firmly entrenched than on land. And love of nature, reverence for the awesome, the elemental, the vital, and the beautiful can be mobilized to reinforce reason and rationality. People everywhere respond to the call, Save Our Seas, intuitively, spontaneously, and it is this kind of response that can create world movements and lead them to success.

The last act of the drama of the oceans is on stage. We must save our oceans if we are to save ourselves.

Pacem in Maribus

BIBLIOGRAPHY

American Society for Oceanography. *The Ocean and the Investor*. n.p.: Dean Witter & Co., in cooperation with American Society for Oceanography, Pacific Western Region, 1969.

Bass, George F. *Archaeology Under Water*. London: Thames and Hudson; New York: Frederick A. Praeger, 1966.

Bates, Marston. *The Forest and the Sea*. New York: Random House (hardcover), The New American Library (paperback), 1960.

Behrman, Daniel. *The New World of the Oceans*. Boston and Toronto: Little, Brown and Company, 1969.

Berrill, N[orman] J[ohn]. *Journey into Wonder*. New York: Collier Books, 1961.

———— *The Life of the Ocean*. New York and London: McGraw-Hill Book Company, in cooperation with The World Book Encyclopedia, 1966.

———— *You and the Universe*. New York: Dodd, Mead & Company, 1958.

Borgese, E[lisabeth] M[ann], ed. *Pacem in Maribus*. New York: Dodd, Mead & Company, 1972.

———— and Krieger, David, eds. *The Tides of Change*. New York: Mason & Lipscomb Pubs., 1974.

Briggs, Peter. *200,000,000 Years Beneath the Sea*. New York and San Francisco: Holt, Rinehart and Winston, 1971.

Brøgger, A[nton] W[ilhelm] and Shetelig, Haakon. *The Viking Ships* (Katherine John, trans.). New York: Twayne Publishers; London: C. Hurst & Co., 1971.

Bryden, John M. *Tourism and Development*. Cambridge: Cambridge University Press, 1973.

Buzzati-Traverso, A[driano] A., ed. *Perspectives in Marine Biology*. Berkeley and Los Angeles: University of California Press, 1958.

Calder, Nigel. *The Environment Game*. London: Secker & Warburg, 1967.

———— *The Restless Earth*. New York: Viking Press, 1972.

Canby, Courtland. *A History of Ships and Seafaring*. New York: Hawthorn Books, 1963.

Casson, Lionel. *Ships and Seamanship in the Ancient World*. Princeton, N.J.: Princeton University Press, 1971.

Charlier, Roger H. "Harnessing the Energies of the Ocean," *Marine Technology Society Journal*, vol. 3, part 1, May 1969, pp. 13–32; part 2, July 1969, pp. 59–81.

Claus, Jürgen. *Planet Meer*. Cologne: Verlag M. du Mont Schauberg, 1972.

Cloud, Preston, ed. *Adventures in Earth History*. San Francisco: W. H. Freeman and Co., 1970.

Cordell, John C. "Modernization and Marginality," *Oceanus*, vol. 17, Summer 1973, pp. 28–33.

Cousteau, Jacques. *The Ocean World of Jacques Cousteau*. 20 vols. Harry N. Abrams, Inc., 1973.

Cowburn, Philip. *The Warship in History*. New York: The Macmillan Company, 1965.

Critchlow, Keith. *Into the Hidden Environment*. New York: The Viking Press, A Studio Book, 1973.

Deacon, G[eorge] E[dward] R[aven], ed. *Seas, Maps, and Men: An Atlas-History of Man's Exploration of the Oceans* [English edition's subtitle: *An Atlas-History of Man's Explorations of the Deep*]. Garden City, N.Y.: Doubleday & Co., 1962; London: Paul Hamlyn, 1962. 2nd ed., 1968.

Engel, Leonard and the Editors of Time-Life Books. *The Sea*. New York: Time-Life Books, 1961, 1969, 1971; school and library distribution by Silver Burdett Company, Morristown, N.J.

Friedmann, Wolfgang. *The Future of the Oceans*. New York: George Braziller, 1971.

Fuller, R. Buckminster; a conversation with Michael Ben-Eli. "Buckminster Fuller on Cities," *The American Way* [published by American Airways], vol. 6, April 1973, pp. 13–20.

_____ "Geoview: Floating Cities," *World,* vol. 1, December 19, 1972, pp. 40-41.

Gaskell, T[homas] F[rohock]. *Physics of the Earth.* London: Thames and Hudson; New York: Funk & Wagnalls, 1970.

_____ and others. *The Floors of the Oceans.* New York: The Geological Society of America, 1959.

The Geological Society of America, Memoir 67. *Treatise on Marine Ecology and Paleoecology,* vol. 1 (Joel W. Hedpeth, ed.). Prepared under the direction of a Committee of the Division of the Earth Sciences, National Research Council, National Academy of Sciences, Washington, D.C. New York: The Geological Society of America, 1957.

Ginsburg, Norton and others, eds. *The Mediterranean Marine Environment and the Development of the Region Inside Malta.* n.p.: Royal University of Malta, forthcoming.

Gross, M[eredith] Grant. *Oceanography.* Columbus, O.: C. E. Merrill Pub. Co., 1967; 2nd ed., 1971.

Hammond, Allen L. "Manganese Nodules (I): Mineral Resources on the Deep Seabed," *Science,* vol. 183, February 8, 1974, pp. 502-3.

Hardy, Alister. *The Living Stream.* New York: Harper & Row, Publishers, 1965.

_____ *The Open Sea,* 2 vols. London: Collins; Boston: Houghton Mifflin Company, 1956, 1959; Boston: Houghton Mifflin Company, one-vol. ed., 1965.

Hawaii, University of, Department of Architecture. "The City and the Sea," *Floating Marine Community Research Report,* by John Craven and others. Honolulu: University of Hawaii for Marine Programs, 1972.

Heezen, Bruce C. and Hollister, Charles D. *The Face of the Deep.* New York and London: Oxford University Press, 1971.

Heronemus, William. "Energy Policies and the International System." Paper for a study project of the Center for the Study of Democratic Institutions, 1974.

Hess, H. H. "History of Ocean Basins." In The Geological Society of America. *Petrologic Studies: A Volume in Honor of A. F. Buddington* (A. E. J. Engel and others, eds.), pp. 599-620. N.p.: The Geological Society of America, 1962.

Hickman, Cleveland P. *Biology of the Invertebrates.* St. Louis: C. V. Mosby Co., 1967.

Hirdman, Sven. *Prospects for Arms Control in the Ocean* (SIPRI [Stockholm International Peace Research Institute] Research Report, no. 7). Stockholm: Almqvist & Wiksell; New York: Humanities Press, 1972.

Holt, S[idney] J. "The Food Resources of the Ocean," *Scientific American,* vol. 221, September 1969, pp. 178-82, 187-94.

Hornell, James. *Water Transport.* Cambridge: At the University Press, 1946; reprint ed., Pomfret, Vt.: David & Charles, 1970.

Horsfield, Brenda and Stone, Peter Bennet. *The Great Ocean Business.* London and Aukland: Hodder and Stoughton; New York: Coward, McCann & Geoghegan, 1972.

Idyll, C[larence] P., ed. *The Science of the Sea: A History of Oceanography.* New York: Thomas Y. Crowell Company, 1969; London: Nelson, 1970. Revised edition published as *Exploring the Ocean World: A History of Oceanography.* New York: Thomas Y. Crowell Company, 1972.

International Conference on Pollution Control in the Marine Industries, 2nd, Washington, 1972. *Pollution Control in the Marine Industries, Proceedings* (Thomas F. P. Sullivan, ed.). Washington: International Association for Pollution Control, 1972.

Isaacs, John D. "The Nature of Ocean Life," *Scientific American,* vol. 221, September 1969, pp. 146-60, 162.

Jehl, Joseph R., Jr. "A Wonderful Bird *Was* the Pelican," *Oceans Magazine,* vol. 2, September–October 1969, pp. 10-19.

Jones, Erin Bain. *Law of the Sea.* Dallas: Southern Methodist University Press, 1972.

Jones, O[wen] A[rthur] and Endean, R., eds. *Biology and Geology of Coral Reefs,* vol. 1. New York and London: Academic Press, 1973.

Kaplan, Irving. "Mater Omnium: Automated Energy and Material Wealth from the Sea," in Borgese, E[lisabeth] M[ann] and Krieger, David, eds. *The Tides of Change.* New York: Mason & Lipscomb Pubs., 1974.

Ketchum, Bostwick H., ed. *The Water's Edge: Critical Problems of the Coastal Zone.* Cambridge, Mass., and London: The MIT Press, 1972.

Landström, Björn. *Ships of the Pharoahs.* Garden City, N.Y.: Doubleday & Company, Inc., 1970.

Linklater, Eric. *The Voyage of the Challenger*. London: John Murray; Garden City, N.Y.: Doubleday & Company, Inc., 1972.

Loftas, Tony. *The Last Resource*. London: Hamish Hamilton, 1969.

Logue, John J., ed. *The Fate of the Oceans*. Villanova, Pa.: Villanova University Press for the World Order Research Institute, 1972.

McConnaughey, Bayard H[arlow]. *Introduction to Marine Biology*. St. Louis: The C. V. Mosby Co., 1970; 2nd ed., 1974.

MacGinitie, G[eorge] E[ber] and MacGinitie, Nettie. *Natural History of Marine Animals*. New York: McGraw-Hill Book Company, 1949; 2nd ed., 1968.

Marcus, G[eoffrey] J[ules]. *A Naval History of England*, vol. 1: *The Formative Centuries*. London and New York: Longmans; Boston and Toronto: Little, Brown and Company, An Atlantic Monthly Press Book, 1961.

Matthews, L[eonard] H[arrison] and others. *The Whale*. New York: Simon and Schuster, 1968.

Mero, John L. *The Mineral Resources of the Sea*. Amsterdam and New York: Elsevier Publishing Company, 1965.

Miller, Robert C. *The Sea*. New York: Random House, A Chanticleer Press Edition, 1966.

Moore, J[ames] Robert, comp. *Oceanography: Readings from* Scientific American. San Francisco: W. H. Freeman, 1971.

Murphy, Robert Cushman. *A Dead Whale or a Stove Boat: Cruise of Daisy in the Atlantic Ocean June 1912–May 1913*. Boston: Houghton Mifflin Company, 1967.

_____ *Logbook for Grace: Whaling Brig Daisy, 1912–1913*. New York: The Macmillan Company, 1947.

Needham, Joseph, with the collaboration of Wang Ling and Lu-Gwei-Djen. *Science and Civilisation in China*, vol. 4, pts. 2 and 3. Cambridge: Cambridge University Press, 1965, 1971.

Nettleton, Arthur. "Cities in the Sea," *Oceans Magazine*, vol. 5, March-April 1972, pp. 71–75.

Newell, Norman D. "The Evolution of Reefs," *Scientific American*, vol. 226, June 1972, pp. 54–65.

Oda, Shigeru. *The International Law of the Ocean Development: Basic Documents*. Leiden: Sijthoff, 1972.

Pacem in Maribus Convocation, Valletta, Malta, 1970. *Pacem in Maribus;* vol. 4 first published by the Center for the Study of Democratic Institutions, 1970. Valletta, Malta: Royal University of Malta Press, 1971, 1975.

Pell, Claiborne, with Harold Leland Goodwin. *Challenge of the Seven Seas*. New York: William Morrow & Company, 1966.

Peterson, Mendel. "Underwater Archaeology," in Idyll, C[larence] P., ed. *The Science of the Sea: A History of Oceanography*, pp. 196–231. New York: Thomas Y. Crowell Company, 1969; London: Nelson, 1970. Also pp. 196–231 in rev. ed. *Exploring the Ocean World: A History of Oceanography*, New York: Thomas Y. Crowell Company, 1972.

Phillips-Birt, Douglas. *A History of Seamanship*. London: Allen and Unwin; Garden City, N.Y.: Doubleday & Company, 1971.

Piccard, Jacques and Dietz, Robert S. *Seven Miles Down: The Story of the Bathyscaph "Trieste."* New York: G. P. Putnam's Sons, 1961.

Pirie, R[obert] Gordon, ed. and comp. *Oceanography: Contemporary Readings in Ocean Sciences*. New York: Oxford University Press, 1973.

Ritchie-Calder, Lord [Peter]. *The Pollution of the Mediterranean Sea*. Bern: Herbert Lang, 1972.

Robertson, Ross. "Sex Changes Under the Waves," *New Scientist,* vol. 58, May 31, 1973, pp. 538–40.

Rougerie, Jacques and others. *Thalassopolis, centres internationaux de recherche de gestion et de controle du patrimoine marin*. Paris: Institut de l'Environnement, 1973.

Ryther, John H. "Food Resources of the Sea," *The American Academy of Arts and Sciences, Bulletin,* vol. 27, February 1974, pp. 15–25.

Sea Island Project: The Building of Islands in the Open Sea Offers Possibilities for Industrial Development. Bos Kalis Westminster Dredging Group N.V., June 1972.

Severson, John. *Modern Surfing Around the World*. Garden City, N.Y.: Doubleday & Company, 1964.

Shepard, Francis P. *Submarine Geology*. 3rd ed. New York: Harper & Row, 1973.

_____ and Dill, R[obert] F. *Submarine Canyons and Other Sea Valleys*. Chicago: Rand McNally & Company, 1966.

Silverberg, Robert. *Sunken History*. Philadelphia and New York: Chilton Books, 1963.

Skempton, A. W. "Canals and River Navigation Before

1750," In Singer, Charles and others, eds. *A History of Technology,* vol. 3, pp. 438–70. Oxford: At the Clarendon Press, 1957.

———— "The Engineers of the English River Navigations, 1620–1760," *The Newcomen Society for the Study of the History of Engineering and Technology, Transactions,* vol. 29, 1953–54 and 1954–55, pp. 25–54.

Soleri, Paolo. *Arcology: The City in the Image of Man.* Cambridge, Mass., and London: The MIT Press, 1969.

Sverdrup, H[arald] U[lrik] and others. *The Oceans: Their Physics, Chemistry, and General Biology.* New York: Prentice-Hall, 1942.

Swann, Mark. "Sea Thermal Power." Unpublished.

Sweeney, James B. *A Pictorial History of Oceanographic Submersibles.* New York: Crown Publishers, 1970.

Toynbee, Arnold J. *A Study of History,* 12 vols. Vol. 1–6, n.p.: Oxford University Press; London: Humphrey Milford, 1934–61. Vol. 7–12, London and New York: Oxford University Press, 1954–59.

Ui, Jun and others. *Polluted Japan.* Reports by Members of the Kushu-Koza Citizens Movement. Tokyo, 1972.

United Nations. *The Sea: Prevention and Control of Marine Pollution.* Report of the Secretary-General. United Nations Document E/5003, May 7, 1971.

———— *Statement of Ambassador Arvid Pardo.* General Assembly Documents: Provisional A/C 1/PV 1515 and 1516, November 1, 1967.

———— *Uses of the Sea.* Study prepared by the Secretary-General. United Nations Document E/5120, April 28, 1972.

Villiers, Alan and others. *Men[,] Ships[,] and the Sea.* New ed. Washington: National Geographic Society, 1973.

Von Arx, William S. and others. "The Florida Current as a Potential Source of Usable Energy." Xeroxed.

Walford, Lionel A. *Living Resources of the Sea.* New York: The Ronald Press Company, 1958.

Ward, Barbara and Dubos, René. *Only One Earth.* New York: W. W. Norton & Company, 1972.

Warner, Oliver. *Great Sea Battles.* London: Weidenfeld and Nicolson; New York: Macmillan Co., 1963.

Wegener, Alfred Lothar. *The Origin of Continents and Oceans* (John Biram, trans.). 4th German ed. New York: Dover Publications, 1966.

Wenk, Edward, Jr. *The Politics of the Ocean.* Seattle and London: University of Washington Press, 1972.

Whittard, W. F. and Bradshaw, R., eds. *Submarine Geology and Geophysics.* Colston Research Society, 7th Symposium, Proceedings (Colston Papers, no. 17). London: Buttersworth, 1965.

Wilson, Carroll L. and Matthews, William H., eds. *Man's Impact on the Global Environment.* Study of Critical Environmental Problems (SCEP). Cambridge, Mass.: The MIT Press, 1970.

Yamada, Yosei. *Yugyotoru Eshi [Pictures of Whaling].* 1829.

INDEX

PHOTOGRAPHIC CREDITS